AF413537

CRYPTIDS

HOW WE KNOW THEY ARE REAL

CRYPTIDS

HOW WE KNOW THEY ARE REAL

*For Cheryl,
who found a strange beast in the wilderness
and brought it home.*

TABLE OF CONTENTS

The Borderland Between Reality and Unreality

When I was born in 1964, there were no cryptids. The term cryptozoology had been coined only a few years before by the zoologist Ivan T. Sanderson. The term translates from ancient Greek as "the study of hidden animals." It wasn't until 1983 that the term "cryptid" was used in the summer issue of the International Society of Cryptozoology newsletter.

The relative newness of the term cryptozoology points to an interesting historical truth—until relatively modern times, the lines between cryptozoology and traditional zoology were somewhat blurred. Travel between continents was time consuming and difficult. Communication within a single country before the age of telegraphs could take weeks or months. Photography is barely two centuries old, and the use of film to capture living animals in motion came even later. For hundreds of years, most naturalists could only study distant animals through second-hand reports, drawings, and the occasional skeleton or pelt. Sometimes living animals would survive being captured and transported across oceans. Occasionally, adventurous students of the natural world would board ships for distant lands to look at new beasts firsthand. Still, until relatively modern times, a scholar of the natural world might become a respected authority on a creature like the rhinoceros without actually having seen one.

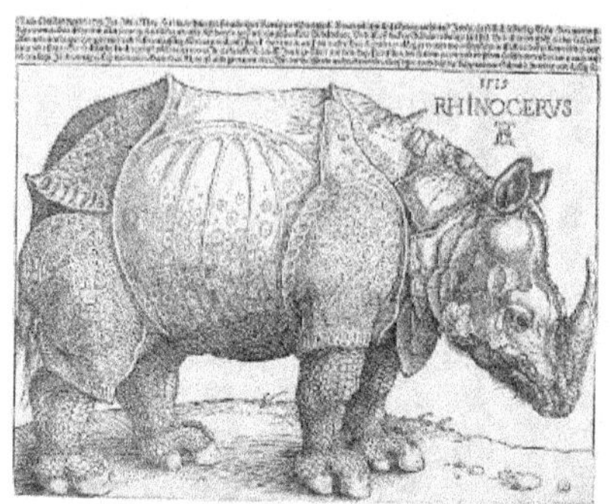

The Rhinoceros by Albrecht Dürer, 1515

There's a European bias to the history of zoology. You can find articles that will tell you that the lowland gorilla was first discovered in 1847; mountain gorillas were discovered in 1902. Of course, the people who lived in the areas of Africa where gorillas dwelled had no doubt been aware of them all along. But the fact that creatures as large and distinctive as the gorilla were still being "discovered" as late as 1902 meant that most naturalists before the twentieth century had open minds. When they heard tales of strange and exotic beasts found on distant continents, or might even plausibly believe that such beasts might be found just beyond the next mountain range. Having seen the bones of mastodons, Thomas Jefferson instructed Lewis and Clark to look for the elusive beasts on their expedition into the uncharted west. Seeking knowledge about hidden and unknown animals was a respectable undertaking when the world was full of barely explored frontiers.

Then, with surprising rapidity, the world shrank. Steam-ships and locomotives sped up travel, letting people cross oceans or continents in weeks instead of months. In the 20th century, airplanes cut those transit times to a few days, or even a few hours. Airplanes were slow compared to radio waves. Drawings of animals found in zoology textbooks gave way to photographs, easily reproduced and distributed in magazines.

The ease and speed of travel combined with the growing accuracy and access to images meant that, by the middle of the 20th century, there simply weren't many places left for large animals to hide. Once impenetrable forests and deserts were crisscrossed by roads and flown over by numerous planes. In 1850, if a zoologist heard about a large, hairy ape lurking in some remote sparsely populated territory, he might organize an expedition. By 1950, similar claims were likely to be met with skepticism, or outright dismissal.

The creatures we commonly call cryptids exist in the hazy borderland between reality and unreality. By definition, they are hidden from direct scientific observation. We have no living specimens currently in zoos, nor have investigations turned up trustworthy physical proof that these creatures exist in the wild. That said, some cryptids are species that definitely, 100 percent existed. We have photos and physical remains of thylacines (also known as Tasmanian tigers) but science has declared them extinct since the 1930s. Yet, some people still claim sightings. Blurry photos and shaky videos of the alleged specimens circulate on the internet. Ambiguous paw prints are occasionally collected. Is this proof the thylacine still exists? Or can we simply dismiss all contemporary reports as hoaxes?

Other cryptids lurk mostly in myth, but with tantalizing traces that they were once real. Most people know what a dragon looks like. They are found throughout multiple cultures, depicted in illustrations and sculptures and literature since the earliest days of human civilization. Few people seriously claim that winged, fire-breathing dragons still lurk in the wild. But, did they ever exist? Some dinosaurs bear at least superficial similarities to dragons. Numerous species of flora and fauna that existed in the age of dinosaurs have survived into modern times. Was there an overlap between a dragon-like dinosaur and modern humans? Or do we concede that people of all eras and cultures have imaginations, and some purely imaginary things can endure without any grounding in reality at all?

Mermaids, for instance, are generally regarded as mythological. But, did numerous cultures independently make them up, or might there be some plausible, grounded explanation for why the idea of aquatic humans is so enduring?

Most cryptids are less concrete than thylacines, but more firmly rooted in the real world than dragons or mermaids. Bigfoot, the Loch Ness Monster, thunderbirds, the chupacabra and Mothman are creatures that have been reported by eyewitnesses. Video and photo evidence sometimes exists, as well as the occasional physical evidence—footprints, mysterious bones, or tufts of unidentifiable hair.

Even with photographic evidence becoming more plentiful than ever, mainstream science doesn't acknowledge the existence of most cryptids. To the layman, what gets accepted and rejected as evidence by science can seem arbitrary. A single footprint imprinted in a 65-million-year-old stone will be regarded by paleontologists as proof of the existence of an unknown dinosaur. At the same time, Bigfoot hunters can collect hundreds of plaster casts of footprints without zoologists accepting these as proof that an unknown, upright ape lurks in North American forests.

We live in a world overflowing with claims that some things are true and some things are false. With so many different opinions, how can anyone be certain of what's real and what's unreal? Experts often disagree. Things commonly accepted as true one day can be universally regarded as false the next day, and vice-versa.

Complicating matters further, we've entered an era where our own senses can't always determine what's real and what's fake. Filmmakers entertain us by showing us realistic sights and sounds of dinosaurs, aliens, and bulletproof women. Such imagery is no longer the exclusive domain of special effects artists. If you have an internet connection and a smartphone, you can produce photorealistic images just by typing a prompt. With that same smartphone, you can edit videos of a politician giving a speech to make it appear that they said something that wasn't actually said. Some people will use this ability to generate and edit sights and sounds to entertain you. Others will use it to scam you. Being unable to tell what's true from what's false can damage your happiness, your health, your politics, and your bank account.

Fortunately, cryptozoology, with its mix of rare but verifiable physical evidence, abundant circumstantial evidence, cultural and historical evidence, false leads, mistaken interpretations, and outright fraud and hoaxes, provides a fertile testing ground for telling absolute truth from absolute fiction. Of course, not everything in the world falls into the realm of absolute truth or falsehood. A great many things must be classified as probable, or unlikely, and the judgment must be made with incomplete and often contradictory evidence.

Studying cryptids sharpens the mental tools needed to cut through the thickets of misleading and incomplete evidence. The skills you'll practice as you judge the existence of Bigfoot can be used to analyze any number of scientific, commercial, political, or philosophical claims. The study of cryptids can lead the open-minded explorer into a better understanding and appreciation of our world.

Depiction of the 1833 Leonids by Adolf Vollmy 1844

This Whole Reality Thing

eality is a nice place to visit, but few of us want to live there.

It seems intuitive that we'd have a strong bias to believe objective, verifiable facts independent of any personal bias. After all, evolution has left us with brains capable of reasoning. Doesn't that mean that, after millions of years of natural selection, we'd be born with an instinctive grasp of what's real and what isn't?

But the first key to grasping reality is simply to observe it. Careful observation of Homo sapiens, either as individuals or as a collective, shows that rationality prevails only occasionally. Most people muddle through life believing numerous falsehoods. Others show a poor grasp of odds. Many people have a fear of flying, which is statistically a far safer mode of travel than the cars they drive with little worry. Many people buy lottery tickets, but see no point in putting four of five bucks into a savings account every week. Yet, over time, the money in the savings account is guaranteed to grow in value, while the lottery ticket buyers are unlikely to cumulatively win more than they've spent.

Why aren't people better at this whole reality thing?

First, a person doesn't need to be 100% correct about everything in life to function as a fairly successful human being. Most people handle the basics of life just fine while believing seven impossible things before breakfast. People embracing numerous falsehoods still find mates, raise kids, and feed, clothe, and shelter themselves. Collectively, irrational and miseducated people manage to form governments. They build villages, cities, and nations. They run schools and churches, and produce useful goods like bricks, bread, copper, computers, sailboats, and satellites. Humanity has obviously thrived despite (or perhaps because of) our delusions.

Second, fitting in has more benefits than simply being right. Most people share the beliefs of their friends and family. You may disagree with your peers on a few details or even a couple of big ideas, but you likely have thousands of shared assumptions and beliefs that you never think twice about. Being a member of a group is an important survival skill. Going along and professing what the rest of your group believes is a behavior that can increase your trustworthiness within the group. You might even adopt and profess differing realities as you navigate through different groups during the course of a day. If you work in a corporate environment, and you're having a discussion with your manager, you might agree with her that the new company reorganization is a clear-sighted path to success. Then, in the breakroom with your fellow coworkers, you might all agree that the reorganization is management covering up their tracks and hiding from accountability for previous poor decisions.

You aren't lying to your coworkers or to your boss. You aren't a hypocrite, you're a human. Why argue about something that you can't change? There's no reward for telling the managers they are wrong. Your coworkers won't love you for correcting their misunderstandings. Maybe it doesn't matter who is right or wrong. Somehow, the world spins on.

Finally, belief in things that are untrue can be valuable. For most of human history, people who said it was impossible to go to the moon were correct. The daydreamers might have believed it could be done, but until it actually happened, the side of the skeptic was the winning side. Now, we've been putting people into orbit and robots onto other planets for decades. Space travel is possible because some people dared to embrace the impossible.

In a world of pure reason, where risks were carefully considered, the world would be much smaller. Going to sea in a small boat of wood or hide was far more likely to result in drowning than it was in the discovery of new lands. Ignorance can be dangerous. But, ignorance is the fuel of curiosity—if you already knew everything, there would be nothing to be curious about. The path toward truth is necessarily paved with ignorance and falsehoods. To reach the known, you must be willing to plunge into the unknown.

I assume you picked up this book because cryptids interest you. They interest me! I've read dozens of books about Bigfoot and the Loch Ness Monster. I've made the pilgrimage to Point Pleasant, West Virginia to get my selfie with the statue of Mothman. I often go hiking and kayaking in wilderness areas, places where cell phone signals don't reach. I've seen things in the

Mothman meets Maxey

shadows, odd movements from the corner of my eye, curious shapes my brain can't quite untangle, and I wonder... while I'm gazing into shadows, are the shadows gazing back?

Countless people claim to have seen large, hairy, manlike beings in the wilderness. There's hardly a culture on earth that doesn't have legends of wild men lurking nearby, sometimes in forests, sometimes underground. Currently, scientists place little value in these claims. But, scientists can change their minds, with enough evidence.

Centuries ago, farmers would occasionally report lights streaking through the sky, followed by a loud boom. Sometimes, in the aftermath they'd find rocks too hot to touch smoldering in their freshly plowed field. For the most part, scientists didn't believe these stories. What kind of yahoo believed there were big rocks floating in the sky? It was far more plausible that lighting had struck a field, blasted away dirt, and revealed a rock. Plus, people had a financial incentive to lie about the sky rocks. People would come from all over the county to look at strange stones. As long as crowds were coming to the farm, a farmer couldn't be blamed for trying to sell a little cider, could he?

There was every reason for skepticism. Sightings were rare, the evidence was ambiguous and easily faked, and there's bountiful examples of people lying about miraculous things to make a few bucks. The only thing the pro space-rock side had going for it was that, in fact, the sky is absolutely swarming with rocks, billions and trillions of them, and thousands of them fall to earth every year. Enough people studied these stones, examining them under microscopes, dissolving them in acid, to prove that the minerals in these rocks were different from other rocks. Once dismissed as mistaken identification and outright fraud, scientists now use meteorites to gather information about the interior of distant stars, the origins of life, and mass extinctions.

The change from presumed myth to established reality seldom happens all at once. Meteorites once belonged to the realm of fakes and folklore. Over time, evidence built and theories developed. What was once regarded as fantasy creeps little by little into the realm of the plausible.

Obviously, there must be some rocks in the sky. I mean, the moon is right there, and it's just a big rock. Still, isn't it odd that there's never an esteemed astronomer on hand when these rocks land? The rocks farmers find have to be hoaxes. I mean, most of them. Some of them, at least. Maybe one in a thousand is really a rock that fell from the sky. Maybe one in a hundred.

Finally, the evidence for meteorites reached a tipping point, and everyone concluded that they were real. More accurately, enough reputable people believed in the reality of meteorites that everyone else went along and accepted that they were real without bothering to examine the evidence themselves.

Will Bigfoot and his cryptid cohort follow a similar path to scientific legitimacy? Maybe. In the pages that follow, we'll be looking at the various lines of evidence for numerous cryptids, and making a good faith analysis of where they lie on the scale between fiction and fact.

Forged in the Heart of Long Dead Stars

The Reality Meter

Studying space rocks in 1750 had one huge advantage over studying cryptids today. If a scientist was willing to hop onto a horse, they could go out and actually look at the rocks while they were still sitting in a crater. Chemistry was a nascent field, with the periodic table of elements over a century away, but an observant scientist could plainly see that these so-called sky rocks were different from rocks commonly found on earth. They might be made of the same minerals, but they were in different ratios. Finding a random earth rock that's a nearly pure blend of iron and nickel is almost impossible. The reality of something you can hold in your hand, or study under a microscope, requires explanation. Is this a hoax? Some misplaced, random bit of slag from a foundry? Or evidence for some new truth, wondrous and illuminating? These meteorites were one piece of a puzzle that led to a profound realization: Everything that makes up our world, including us, is assembled from elements forged in the hearts of long dead stars.

Cryptids, unfortunately, aren't as easily studied as meteorites. Once a meteor falls from the sky, it more or less sits there until someone picks it up. Cryptids are, presumably, living beings. They don't wait patiently in one spot waiting to be discovered. Of course, zoologist capture living creatures all the time, and dead specimens of various species can be viewed in museums and universities. With a few noteworthy exceptions we'll examine in later chapters, the chief obstacle to studying most cryptids is that we don't actually have bodies to examine. We seldom even have body parts, like a skull or a pelt. We rarely even have scat! For quality evidence of a creature's physical reality, a single pile of poop is more valuable than a hundred eye-witness sightings.

On the other hand, simply because scientists haven't yet collected a specimen of an animal isn't proof that the animal doesn't exist. The world is a big place. Scientists go out into the field every year and collect bugs, flowers, salamanders, and even mammals that no one has cataloged before. In 2012, a new species of monkey was discovered in the Congo, the lesula. Of course, locals had known about the monkeys. They didn't need a biologist to confirm that lesulas were real. But, until biologists went and documented the animals, a lesula was no more likely to turn up in a zoology text book than Bigfoot.

Just as natives of the Congo likely knew about the lesula, indigenous Americans believed that large, hairy men shared the forests with them. This is documented in part by pictographs, like the Hairy Man drawing dating from about 500 AD found in the San Joaquin Valley in California. Even if there are no surviving members of this species of Hairy Man, perhaps such creatures existed contemporaneously with early humans in the area, before being driven to extinction? We know from fossil records that humanlike species such as Neanderthals and Gigantopithecus existed in Asia in the not so distant geological past. If humans could walk across land bridges to North America during an ice age, there's no logical reason Gigantopithecus couldn't make the journey. These apes could be nine feet tall. It's a safe bet they left behind very big footprints!

Alas, while it's possible that Gigantopithecus could have beaten modern humans to North America, the argument that they did has one major weakness: a lack of any physical evidence. On the other hand, the only evidence that Gigantopithecus existed in Asia comes from fewer than twenty partial specimens. Preserving remains over thousands of years involves a great deal of luck. For every species we identify in the fossil record, there's multitudes of creatures that lived in habitats where they were unlikely to be fossilized. If they were fossilized, it's possible we haven't found them. If we have found them, it's possible we didn't realize what we'd found, and the remains might be hidden in some crate in a university basement, overlooked and long forgotten.

The discovery and verification of species long extinct is tedious, difficult work. Money and media flow to big discoveries. Literally big! Dig up the jawbone of a dinosaur bigger than a Tyrannosaurus Rex and newspapers around the world will talk about your discovery. If you carefully document evidence of a new species of long extinct clam, your scholarly peers might appreciate your efforts, but your discovery is unlikely to be breaking news on CNN.

This bias towards big, flashy discoveries is one reason why cryptids receive a disproportionate degree of media coverage. Discover a new beetle, and the world yawns. Find the decaying body of a yeti that fell off a cliff? Your name is going into history books.

But the fact that fame is a likely reward for cryptid discovery muddies the water. We have numerous examples of people who were hungry for fame and willing to commit hoaxes to get it. Even more people desire fortune, and there's money to be made in cryptids. When I went to Point Pleasant for my selfie with Mothman, I got myself a t-shirt at the museum. Thousands of people have spent money in that gift shop. This doesn't mean the people who run the Mothman museum are guilty of a hoax, nor does it prove Mothman doesn't exist. It does show that people who advocate for various cryptids might have motives other than a heroic devotion to the truth. The fact that some people might pursue fame, and others profit, and others might simply enjoy inventing stories purely for entertainment, means we must take extra care as we examine the evidence for cryptids.

Starting in the next chapter, I'll be examining cryptids one by one and weighing in on how likely they are to be real based on current evidence. Many books on cryptids rely heavily on recounting eyewitness stories of encounters. Very little of this testimony will be found in these pages. Unfortunately, and paradoxically, the more eyewitness accounts there are regarding a cryptid, the less likely it is that the accounts are true.

The more people describe a creature like the Loch Ness Monster, the more eager others will be to share their own stories. Sometimes this will be a simple bias toward mistaken identification. You've heard stories of a monster in the lake. You see, in the distance, black bumps in the water, leaving a wake. Maybe you're just seeing the wind pushing a floating log. But where's the fun of going to the nearest pub and exclaiming, "I just saw a log!"

"I just saw a log! And what were probably birds!"

I'm hesitant to call any specific claim of a cryptid sighting a hoax. Still, it's naïve to think that there aren't numerous hoaxers out there. It's not that hard to fake a footprint in damp sand. It's not difficult to use an infrared camera to fake a vaguely man-shaped blob loping up a hill in the distance.

Once you start studying cryptids, you encounter a glut of information. Tens of thousands of eyewitness accounts. Countless hours of shaky video. Innumerable blurred photos. Can you trust the photos that aren't blurry? Anyone with access to the internet can type prompts into an AI image generator and with a little patience and tweaking produce a passable image of a Bigfoot lurking in a forest. Many of the illustrations in this book were created this way, though I've deliberately made an effort to avoid photo-realism. The last thing I want to do with this book is clutter the world with faked evidence.

If we can't believe eye-witness testimony, and if visual evidence like photos and videos are easily faked, what's left? Have I stacked the deck to simply dismiss all cryptids as fiction?

Of course not. A great deal of reality has yet to be verified by science and we must make the best judgments we can from available information.

Like the rocks that fell from space, cryptids, if they are real, should leave behind evidence we can hold in our hands and subject to a range of scientific tests. Assuming that cryptids are creatures of the natural world, we'd expect them to follow biological laws. That means, among other things, that we aren't looking for one immortal specimen of a given cryptid spotted decade after decade, but breeding populations of multiple individuals. It also means that a cryptid can only exist in a habitat that would provide them sufficient nutrients to survive, and sufficient cover to conceal themselves. Numerous species have been driven to extinction simply because humans have degraded their environment.

Another law of biology is that all creatures descend from other creatures. Any modern organism has precursor organisms in the fossil record, or related species that can still be found. The dogs in your life have links to Asian wolves. For a cryptid to be plausible, it should fit somewhere on the evolutionary tree. (For now, at least. As we move further into our modern age of genetic manipulation, we might one day encounter creatures with no precursors in the fossil record because they've been built from scratch. If the evidence for dragons in our past is unconvincing, we might yet populate the future with toothy, winged beasts.)

I'll note that I lack the credentials of a scientist. But science isn't the private property of scientists. Knowledge belongs to anyone who makes the effort to gain it. If I make a claim about evolution or genetics or climate, I encourage you not to take me as a final authority on the subject. Libraries are filled with books on these topics. Read widely. Visit museums. Go out into forests and fields and take the time to look at the natural world for yourself. All of reality awaits your study. Everything is interconnected. You can't really determine the truth about cryptids until you've gone out and spent time contemplating flowers, fields, and forests and looked at squirrels, frogs, worms, and bugs. You come to see how things fit into the natural order, how everything that exists is linked to everything else. There's no point seeking to study the hidden things of the world, like cryptids, if you aren't equally hungry to discover the things that are right in front of you.

With this in mind, I'll be judging five criteria to score each cryptid for their "realness." Each score will fall on a scale from 0-20, with 0 being absolute lack of evidence, and 20 being indisputable, well-documented evidence. The evidence we'll examine will be:

Physical evidence from modern times: Do we have a body?

Fossil or other preserved evidence: Do we have proof that precursor species existed?

Biological viability: If no direct fossil evidence exists, is there other evidence from similar species that prove that the proposed creature could be real?

Habitat: Are the environmental conditions favorable to supporting such a creature?

Concealment: Would the habitat allow the creature to easily elude detection?

Alas, while a cryptid can earn points in all these categories, there needs to be one more category that subtracts from the overall tally:

Strength of alternative explanations: Are mermaids just manatees? Are chupacabras only feral dogs? If the evidence for the alternate explanation is solid, I'll be subtracting up to 20 points. In some clumsy foreshadowing, if you're a true believer in the Loch Ness Monster, this part of the analysis is going to be rough for you.

I'll be placing all these points on a "Reality Meter." A score of 100 indicates a cryptid that has been verified beyond all doubt. It's as real as an elephant or a hummingbird.

Once the score drops down below 75, a little faith will kick in. The existence of such creatures is certainly not proven, but it's not implausible. If you wake up tomorrow and read that scientists have captured a live specimen, it's not going to shock you.

At around 50, the odds stop being in favor of the cryptid's actual existence. Science can't 100% rule out that a specimen might yet turn up, but the lack of hard evidence means belief in the creature is based more on faith than facts.

Anything with a score lower than 25 is purely for entertainment value. It's like believing in Santa's flying reindeers. It's harmless to believe in them, and arguably even a bit wholesome. But, if a gravity-defying reindeer ever did turn up, it would overturn a great deal of what we think we know about evolution, biology, and physics!

Now that I've explained the rules and the logic behind my analysis, let's see who and what I'm going to analyze:

The Abominable Snowman – Do large, hairy human relatives haunt the frosty reaches of Asia?

Sasquatch – Are North American forests home to elusive, upright apes?

Jackalope – That taxidermy jackrabbit with antelope horns hanging in a bar is an obvious fake… or is it?

Mermaids – Did mutant primates colonize the sea?

Lizard Man – Do the swamps of South Carolina hold a dark secret?

The Loch Ness Monster – Have plesiosaurs survived into modern times?

Mothman – What explains the mysterious phantom of West Virginia?

Thunderbirds – What were these airplane-sized avians?

Dragons – Why do so many human cultures believe in giant flying lizards?

Thylacines – If it's extinct, why am I seeing videos of it on YouTube?

Chupacabra – Have we figured out what's sucking the blood out of all these goats?

Kraken – Have we found the monster cephalopod of yore?

This may look like a surprisingly small list considering the sheer number of cryptids in the world. But, in many cases, I'm picking a single representative cryptid to represent a larger, broader class of such beings in an effort to avoid redundancy. The Loch Ness Monster has parallels in other aquatic monsters like Chessie in the Chesapeake Bay, or Champ of Lake Champlain. Our analysis of thunderbirds can apply to other giant birds like the elephant-nabbing roc. The noteworthy exception will be to examine both the Abominable Snowman and Bigfoot. Though both are large, shaggy man-apes of the wilderness, the supporting evidence for each has its own strengths and weaknesses worthy of examination.

ABOMINABLE SNOWMAN

The Abominable Snowman

efore Bigfoot became a pop culture sensation, the first large, hairy, manlike beast to enter mainstream American awareness was the Abominable Snowman. Henry Newman, a journalist working in Calcutta in the early 20s, heard reports of a man-like creature sometimes spotted in the Himalayas. The locals called the beast *metoh kangmi*, which translates to "manlike wild creature." Newman, looking for a way to make these tales interesting to his readers, hit the thesaurus and decided that *Abominable Snowman* was a catchy translation. He likely didn't anticipate that his reporting would launch a legend still being talked about a century later.

While the distinctive name was no doubt important in planting the Abominable Snowman into public awareness, the timing of Newman's report is also important. In 1921, the world was accustomed to opening newspapers and discovering wonders. Improvements in technology like radio and photography let Information flow across the globe. Giraffes, ostriches and tigers had been displayed in zoos and circuses, but photography allowed people to actually see these creatures in their native habitats. Most importantly, new species of large animals were still routinely documented by zoologists. The mountain gorilla was discovered in 1902. Bonobos, one of our closest evolutionary relatives, weren't identified as a distinct species until 1929. So, in 1921, the idea that a new species of ape was on the verge of discovery felt plausible, or even probable.

Over time, the Tibetan name for the creature, *yeti,* pushed aside the usage of Abominable Snowman. It's nice to think that this was due to Western cultures becoming more respectful to native cultures and adopting their terminology. My more cynical speculation is that the *Rudolph the Red Nose Reindeer* TV special literally pulled the fangs out of the Abominable Snowman, changing him from a savage beast to a cuddly toy kids wanted for Christmas. Rebranding as the yeti was needed to get the wild man-beast of the Himalayas to be taken seriously again.

For a creature that's never been photographed, the image of a yeti has become increasingly clear over time. The beasts are thought to be broader and taller than humans, covered in white hair, with an upright gait. Numerous footprints have been found, quite large, and also quite manlike. If the Abominable Snowman does exist, it's likely that genetic analysis would show it to be close to humans on the evolutionary tree.

Unfortunately, the footprints aren't firm evidence. Himalayan brown bears walking through snow can have their hind paws fall partway into the print formed by the forward paw. The resulting overlap vaguely resembles a giant human footprint. Alternatively, footprints in snow can grow with melting and refreezing. If you live in an area where snow is common, you've likely seen this happen with your own footprints.

As mentioned in the opening chapters, the various eyewitness accounts can't be taken as hard evidence. Eyewitness accounts are only useful if they can lead us to some sort of physical evidence to verify them.

How about direct physical evidence? Yeti hides and hairs are occasionally presented for analysis. Most have been identified as belonging to more mundane creatures. The most famous relic, a supposed yeti scalp, turned out to have the DNA of a Himalayan brown bear.

But, the lack of evidence isn't proof that the yeti, or a creature very much like it, can't exist. The biggest reason not to dismiss the yeti out of hand is that it lives in Asia. The Eurasian continent is thick with legends of hairy men of the wilderness. Almas are found in Turkish, Kazak, and Mongolian folklore. Pagan traditions throughout Europe tell of the Green Man, a humanoid creature that embodied the spirit of the forest. A Neanderthal-like wild man known as the chuchunya is said to be found in the forest of Siberia.

What's intriguing about these legends is that science has established quite firmly that there were, in fact, very close relatives of humans living in parts of Europe and Asia. The Neanderthals are the most well-known, but Denisovians also overlap with modern humans. Their remains are found in Siberia. There aren't firm dates for when they died out, since Denisovian remains weren't identified until 2010. They were mistaken for Neanderthals at first, but DNA testing showed they were a genetically distinct species.

Another interesting distant human relative was the Gigantopithecus, whose remains are found in China. It was more gorilla-like than human, and probably walked on all fours. But, it was big, larger than modern gorillas, and likely quite imposing when it stood erect. I'm using qualifiers like *probably* and *likely* because, unfortunately, only partial remains have ever been found.

Still, no one disputes that Gigantopithecus was a real creature. For any remains at all to have been found, there must have been a large enough population to sustain breeding. They also couldn't have evolved in utter isolation. It's also probable there were related species throughout the region, whose remains will never be found simply because the odds of the remains of any extinct species being preserved are quite low.

Asia was indisputably populated by humanlike creatures as recently as 40,000 years ago. In geological terms, this is quite recent. It's wildly improbable that the remains we've found were left by the last living members of these species. Smaller groups we haven't found evidence of must have carried on for centuries or even thousands of years. Are tales of the yeti and its Asian and European relatives the result of remnant populations of Denisovians, Neanderthals, or Gigantopithecus? Could the need to avoid Homo sapiens have driven these creatures into increasingly remote areas where humans seldom travelled?

If a species of human-like ape has survived into modern times, evolutionary pressure would likely make it shy and elusive. This could explain why they are rarely encountered. Of course, while this speculation is informed by what we know about evolution, it is still only speculation. We have no living member of our mystery species to examine to determine its actual habits.

Given these facts, just how real is the Abominable Snowman? Or, to frame the question as broadly as possible, how realistic is it that some other large, near-human relative is surviving today in some remote Asian wilderness?

Physical evidence from modern times: This is the weakest link in proving the existence of the Abominable Snowman. If yetis were legitimately leaving behind so many footprints, we'd almost certainly be finding hairs, scat, or even corpses. Dogs would catch their scent. The one reason to think that the earliest footprints to be photographed might be legitimate is that they were taken in good faith by explorers who gone out to climb mountains and had no interest in finding yetis. These early footprints could be misidentified, but they aren't likely to be hoaxes. Being generous, out of 20 points, these few footprints earn our Asian man-ape 4 points.

Fossil or other preserved evidence: A solid, easy 20, as long as we're applying a broad definition of "man-like creature." There's fossil evidence of near-human relatives all throughout Europe and Asia, some of it of geologically recent vintage.

Biological viability: Again, an easy 20. Go to your nearest natural history museum and you can no doubt gaze upon the bones and skulls of Neanderthals or some other early human relative. Early man shared the world with numerous related species.

Habitat Strength: Our early ancestors and related species showed a remarkable ability to reach distant lands and exploit the resources found there. Humans thrive in numerous habitats, from remote jungle islands to frozen tundra to blazing deserts. Our nearest relatives often settled these terrains before we did. While much of the natural landscape has been altered or polluted by the spread of human civilization, the Eurasian landmass is immense and full of viable niches where a large, apelike creature might survive. We can again give this a maximum score of 20.

Concealment. Alas, due to sheer population density in Europe, it's highly unlikely that any wild men remain in the woods there. On the other hand, Siberia, where the Denisovian remains are found, is one of the least densely populated areas on earth. True, Siberia is quite distant from the Himalayas, so any creature living there wouldn't technically be the Abominable Snowman first reported in 1921. But, if we found any living man-like species anywhere in Asia, it would be evidence that other varieties of Asian wild-men have their roots in fact. Siberia is a strong candidate for a habitat where these creatures could still hide. It has over a million square miles of dense forest with few roads. Avid cryptid hunters can't drive up for a weekend and set up a few trail cams. The thick forest likely makes surveillance by drones nearly impossible. Dogs might be able to sniff out a strange beast, but, given that humans and dogs have been closely associated with each other for tens of thousands of years, it's not difficult to theorize that an elusive ape might be good at catching the scent of dogs and staying well clear of them. If a shy humanoid wants to elude mankind, this is one of the last wild places on Earth where it's still plausible. However, since it's pure speculation that these creatures would evolve to be elusive, or evolve to avoid dogs, I can't score this a 20. At best I can give our hidden man-ape a 15.

So far, we're at a score of 79. Not bad! But, now we need to look at the points we should subtract:

Strength of alternative explanations: The most common explanation for sighting the yeti or its relatives is that people are seeing bears. I'm dubious that this explains more than a few sightings. I spend a lot of time in remote areas, hiking, biking, and kayaking. It's very rare to encounter them, but bears do occupy these areas. One bike trail in Virginia had explicit instructions on what to do if you encountered a bear on a trail. The few times I've caught a glimpse of a bear, usually at a distance and in retreat from me, I've had no confusion at all what I was seeing. It seems insulting to think that natives of Asian lands or experienced mountaineers would confuse a bear for a man-ape, even if they saw it walking on its hind legs. People are good at identifying bears! People are even better at spotting other people. This… could be a problem.

Can you see a face in this photo of a Martian hill?

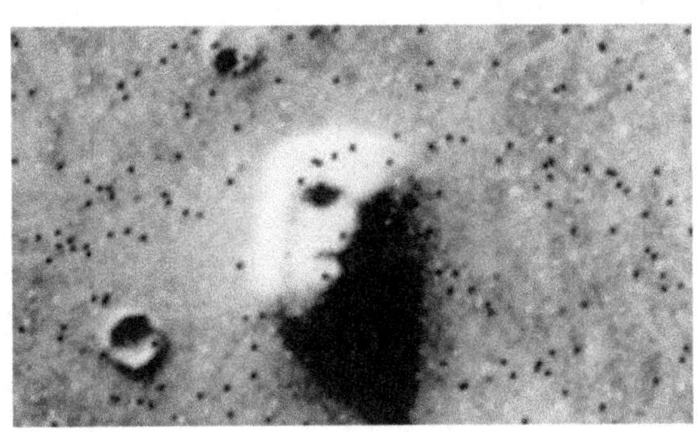

How about in this electrical outlet?

Humans have a gift for seeing faces, even where none exist. The scientific name for this is pareidolia. This is almost certainly a product of evolution. We're social creatures. Quickly identifying and reading the faces of people around us to judge their moods and needs is important to social cohesion of groups.

A second evolutionary driver of our face-spotting skills comes from the dark fact that the dangerous animal most humans are likely to encounter another human. Through hundreds of thousands of years of evolutionary history, "other humans" was a big category that included species like the Neanderthals. Being able to spot a manlike shape in the shadows of a forest is an important survival adaptation.

Many people, out alone in wild places far removed from other humans, have experienced the eerie feeling that they are being watched. This instinct is there to keep us safe. When you're alone, in wild spaces far removed from other humans, you're at your most vulnerable. Your eyes flicker across the surroundings, looking for any potential enemies. A false signal, where you think you see a menacing face looking at you from a distance, bears little evolutionary cost. You skedaddle back to your fellow tribesmen, safe and sound and with a story to tell. On the other hand, if there is a face of a hostile humanoid peeking through the branches, and you failed to see it, you might never make it home.

Ultimately, I suspect this evolutionary hardwiring to be on hyper-alert when you are in remote areas is a contributing factor to the various hairy wild-man sightings. The evolutionary echo of competing with near-human species gives a blurry mental picture of what we need to fear.

Pareidolia is a well-studied and proven quirk of human perception. It likely can't explain all sightings of creatures like the yeti, but it certainly explains some of them. For this, I'll subtract 10 points.

Subtracting that out leaves the yeti with a score of 69. It's a long way from being established as a creature that lives in the modern world, but it's also well-removed from the realm of pure fantasy. The physical evidence of similar creatures living beside modern humans into geologically recent times, combined with a vast potential habitat, makes further searches for a hairy human relative living in the Asian wilderness a worthy pursuit.

The Reality Meter
The Abominable Snowman

Modern Physical Evidence	4
Fossil Evidence	20
Biological Viability	20
Habitat Availability	20
Concealment	15
Alternative Explanations	-10

BIGFOOT

Bigfoot

Bigfoot is arguably the most famous of all cryptids. Sasquatch is another common name for the creature, used more or less synonymously. Outside of the Pacific Northwest, there are local variants of Bigfoot like the Skunk Ape in Florida, and the Boggy Creek monster of Arkansas. Any given version of the beast has its champions, but for our analysis we'll be casting a wide net. Is there a large, bipedal, non-human ape hiding somewhere in the wilderness of North America?

The name Bigfoot became popular in 1958. Loggers in Humboldt County, California, started telling people about the large, humanlike footprints they were discovering as they cleared the forest. This was reported by a local paper, then the story was picked up by outlets like the Los Angeles Times, and soon spread nationwide. Since then, thousands of sightings have been reported. Numerous plaster casts of purported footprints have been made. The original range of Bigfoot is the Pacific Northwest, but has grown to include sightings in nearly any semi-remote, forested areas, leading to Bigfoot festivals as far east as Marion, North Carolina.

Does the rapid spread of sightings indicate people talking about creatures that they'd been aware of all along and reluctant to talk about? Or is it a social contagion, where an idea gets put into people's head, is popularized by the media, and eventually takes on a life of its own?

Browse casually through television channels and you have a good chance of encountering Bigfoot and his relatives. He's been the subject of feature films. He shows up in shows for children. The *History, Discovery,* and *SciFi* channels all have numerous documentaries about the search to find the elusive man-ape. Bigfoot appears in commercials selling beef jerky and car insurance. You find his image on shirts and coffee mugs. Plaster casts of Bigfoot's big foot

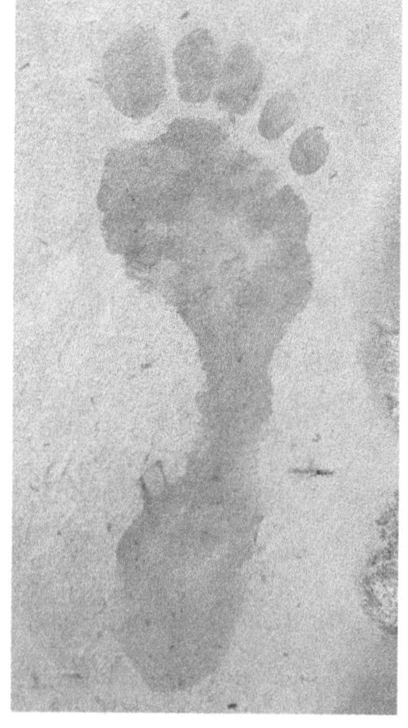

This + plaster = $

are bought and sold on Amazon. There's even a monster truck named after him.

This prominence in popular culture makes trusting contemporary accounts of Bigfoot a little tricky. Anyone raised in America in the modern era can likely tell you what Bigfoot looks like by the time they reach elementary school. There are life size plywood cutouts of Bigfoot that people place in their yards or in front of their business. No identifying placard is needed; people can identify Bigfoot simply by his silhouette.

Bigfoot's popularity means, unfortunately, that truly neutral eyewitness encounters of the creature are non-existent. People's ideas of what Bigfoot should look like are well formed at an early age, so the consistency of reports carries little weight as evidence. Video and photos are suspect as well. People's foreknowledge of what Bigfoot looks like makes him an easy target for hoaxes. The most famous Bigfoot video evidence, the Patterson–Gimlin film from 1967, is a short snippet of Bigfoot casually strolling past the camera. It only lasts a few seconds. It's either a genuine Bigfoot, or a guy in a costume. We have plentiful evidence that people can make and wear costumes. Unfortunately, the costume explanation is more in agreement with mundane reality.

The Patterson-Gimlin film was made after the story of Bigfoot was already widespread. In fact, Patterson had already published a book on the subject: *Do Abominable Snowmen of America Really Exist?* Patterson wasn't a neutral observer.

Since most modern accounts should be treated skeptically due to bias created by Bigfoot's presence in popular culture, are there reports or images that predate his fame?

One intriguing image is the Hairy Man petrograph, a painting on a rock found at the Tule River Reservation, Tulare County, California. Documented in 1889, the painting is estimated to be a thousand years old. It depicts an entity called Mayak datat by the local tribe, the Yokuts. This translates to Hairy Man, and it's easy for a modern observer to look at it and think, "Yeah, that's Bigfoot!" It's not a slam dunk proof. The paintings and sculptures of cultures worldwide are full of depictions of mythological and spiritual entities. We don't, for example, treat sculptures of Zeus as proof that Zeus actually existed.

Painted Rock, Tulare County, Ca.

That said, cave paintings often depict large mammals that were once abundant but which went extinct before modern times. As a depiction of a tall, hairy man with big feet drawn with no possibility of contamination by modern popular culture, it's a tantalizing clue.

How plausible is it that Bigfoot existed in North America thousands of years ago? Humans are estimated to have first settled North America 20,000 years ago. We also have strong evidence that Neanderthals existed as recently as 40,000 years ago. Viewed through geological times, the 20,000 year mismatch isn't an insurmountable barrier. If Neanderthals were hunted and killed by our species, it's easy to imagine the remaining Neanderthals retreating further and further into the wilderness to avoid mankind. Since things like campfires or long-term

A 1909 illustration of a Neanderthal by Frantisek Kupka. This is more apelike than modern reconstructions, but it's still a plausible Bigfoot.

encampments might reveal their location to humans, they might have learned to avoid these things. Once they stopped making campfires or staying in one place, they would mostly vanish from the archeological record.

It's easy to imagine some remnant population of these Neanderthals retreating across the land bridge to North America to avoid humans in Asia. Then, alas, humans followed them, grabbing the best and most accessible lands by rivers and coasts. The human-adverse Neanderthals would have had no choice but to retreat to rugged mountains and dense, tangled forests to elude their most dangerous predators. Evolutionary pressures would encourage strong avoidance instincts toward man, and evolution could also drive sensory enhancements that could give these remnant Neanderthals the ability to smell, hear, and see us before we're ever aware of them. Perhaps they lived on the border of indigenous American society for thousands of years before going extinct in relatively modern times. Or, maybe they are still out there, shy and reclusive, forever one step ahead of a species it has every reason to fear.

Personally, I think this is a pretty good argument. The one teensy flaw in the argument, alas, is any physical evidence whatsoever.

Other animals possess similar strong instincts to avoid humans. But, biologists and wildlife photographers routinely track down elusive beasts. Trail cams capture images of pumas, bears and wolves, all possessing sharp senses that make them difficult to find. Even if his trail cams never caught a picture of a bear, a hunter will have no trouble telling if one is in the area. Animals create and follow trails back and forth to spots where water, food, and salt licks can be found. Follow one of these trails in a bear infested area, and you will find scat. If you spend much time in the wild, you're also likely to stumble across bones. Bushes and bark will hold onto tangled bits of hair. Real animals produce a massive amount of biological debris.

If large, hairy, manlike apes were genuinely being spotted as often as reported, we'd find stronger physical evidence than footprints. So far, no scat. No bones. Of course, Neanderthals were known to bury their dead. Maybe Bigfoot does the same. But, this creates even more difficulties for the lack of evidence, since burials preserve bones longer than simply leaving corpses scattered about in the open, where they're more swiftly disposed of by carrion birds and beetles.

As we score Bigfoot on our reality meter, things start off grim.

Physical evidence from modern times: A Bigfoot skull was recently reported, but it was instantly dismissed as fake. No hair or scat unmistakably from an ape-like species have yet been recovered. The footprints are numerous, but mostly date from decades following photographs of the supposed footprints of the Abominable Snowman. With nearly any other animal, if you can locate prints, you won't be far from evidence like hair or scat. Unfortunately, I don't see how physical evidence will get a score higher than 1.

Fossil evidence or other preserved evidence: If we're feeling generous, and interpret the Hairy Man petrograph as a depiction of a real creature the way we interpret European cave paintings as depicting now extinct animals, we can count this as having some modest weight. Unfortunately, North America simply lacks fossil evidence of any species closely related to humans. It's true there are fossils in Asia, and it's true that Asia and North American have been connected in the past. But, in the absence of even a single humanlike fossil in North America, we'll need to score this a 2.

Biological plausibility: Here we have the Bigfoot's strongest claim to reality. While North America has little evidence of such creatures, we find the bones of human-related species throughout Asia. We know that Asia and North America have been linked by land bridges during ice ages, giving them a path here. This earns a 20.

Habitat strength: The mountains of the Pacific Northwest are more than capable of supporting large mammals. If a grizzly bear can find enough food, Sasquatch should be able to rustle up some vittles. Even more to the point, our own ancestors found the primitive North American landscape to be welcoming territory, and successfully spread to its most remote corners in only a few thousand years. Another 20.

Concealment: While it's plausible that a Sasquatch could survive, it's less likely that they could elude detection. It's worth noting that the modern encounter that launched the Bigfoot's popularity was an encounter with a logging crew. The forests are vast, but since people first arrived in the area they've been exploiting the forests for natural resources. Trails crisscross the mountain ranges, and modern inventions like trail cams, infrared cameras, drones, and ubiquitous cell phones make eluding detection harder than ever. Finally, at the risk of seeming obsessed with poop, the lack of Bigfoot scat is difficult to overlook. It varies from animal to animal, but living creatures leave behind a lot of bodily waste. An average human produces twice his or her own weight in poop over the course of a year. If the same ratio applied to a Bigfoot, that's a lot of crap. Because the American wilderness is better explored and more heavily exploited than the Siberian Taiga, I feel that at best I can score this a 10.

Bigfoot earns a score of 53, and we haven't even subtracted out the score given for the **strength of alternative explanations.** We face the same likely explanation that complicates sightings of a yeti, pareidolia. Our brains are simply programmed to see humans and human features where none exist. Once again, we'll need to subtract 10 points.

The case for a North American large, hairy, bipedal ape looks shaky. It's not 100% impossible. No biological laws would be upended if a hiker stumbled across a Bigfoot corpse tomorrow, giving us a clear, indisputable specimen. But, on the reality scale, with a score of only 43, Bigfoot is likely only an entertaining myth.

The Reality Meter
Bigfoot

Modern Physical Evidence	1
Fossil Evidence	2
Biological Viability	20
Habitat	20
Concealment	10
Alternative Explanations	-10

JACKALOPE

Jackalopes

ackalopes are creatures with the body of a jackrabbit and the horns of an antelope. Whether they are a distinct species, or the result of some unlikely hybrid mating between two species that science says could never produce offspring is open to debate. But, that debate isn't one we need to get into here, since better alternatives to either explanation immediately jump forward.

At first glance, a skeptical cryptid hunter might be tempted to dismiss the jackalope as an obvious hoax. Starting in the 1930s, numerous stuffed jackalopes started appearing in barrooms and hotel lobbies. The legend of the jackalope predated the appearance of the mounted specimens, however. Horned rabbits were commonly believed to exist in Europe, and an illustration of a creature similar to a jackalope appears in *Animalia Qvadrvpedia et Reptilia (Terra)* by Joris Hoefnagel, a bestiary dated to roughly 1575.

Setting aside this respectable historical lineage, the origins of the popular modern depiction of the jackalope are well established. They were made by a taxidermist

Animalia Qvadrvpedia et Reptilia (Terra) by Joris Hoefnagel. 1575

named Douglas Herrick who lived in Douglas, Wyoming. Herrick made his jackalope from the hide of a jackrabbit and the antlers of a deer. (Despite the name being a combination of jackrabbit and antelope, the vast majority of depictions portray the creature with deer antlers as opposed to antelope horns. Antlers branch; antelope horns don't.) The creation was put on display at Douglas's La Bonte Hotel in the early 1930s, where it became a tourist attraction. The Douglas Chamber of Commerce sold licenses to hunt jackalopes. The horned rabbit's fame has only grown since, and typing the word into Amazon produces over a thousand results for toys, hats, books, shirts, and "authentic" mounted jackalope heads you can proudly display in your own home.

Here we run into an interesting paradox of this particular cryptid. With most cryptids, eye-witness sightings are common but preserved remains are non-existent. The jackalope flips this script. Without much effort, you can probably find a stuffed jackalope on display somewhere within driving distance of your house. But, stories of close encounters with creatures in the wild are relatively rare compared to a cryptid like Bigfoot. You don't have multi-season runs of cable TV shows searching for the beast. Even hardcore cryptid fans look upon the critter and suspect there's simply nothing there worth searching for.

Except… horned rabbits are totally a thing.

It's true, they seldom resemble the classical image of a jackalope. But, rabbits are vulnerable to infection by the Shope papilloma virus. This virus causes tumors on the hides of rabbits and hairs, and the tumors are made up of keratin—the substance that makes up nails, hair, and horns. These tumors can sometimes grow quite large, producing horny growths several inches long. These can be located anywhere on a hare's body, but specimens of these horns growing from a rabbit's head are well documented. Unfortunately, this doesn't produce cool antlers on a healthy rabbit people would be eager to display as a trophy. Photos of rabbits with these horny tumors jutting from their jaws or foreheads are rather sad looking. You can google them with ease, but I'll refrain from reproducing them.

Jackalopes might not exist as commonly portrayed, and they may not be a separate species, but horned rabbits are indisputably genuine. Therefore, using an admittedly broad and generous interpretation, I feel comfortable declaring that the jackalope is an animal that could, in fact, be encountered in the wild. Random chance must occasionally produce twin keratinous tumors on the heads of rabbits that give us a specimen close enough to the popular image of a jackalope. The real life existence of horned rabbits also explains why the creature can be found depicted in bestiaries drawn four centuries before some taxidermist in Wyoming felt creative.

The realization that something like jackalopes exists in nature has an interesting historical parallel in the platypus. When preserved specimens of the platypus first started showing up in Europe, skeptics assumed they were taxidermy hoaxes. Here was a creature that seemed to be half duck. It had a bill and laid eggs. But, it also had fur, and a beaver's tail. Unlike either birds or mammals, it also possessed venomous spikes on its ankles!

To accept the creature as genuine would require rewriting the laws of biology. On the other hand, a clever taxidermist could slap together the body parts of different animals to produce all sorts of fantastic critters. Genetic testing wasn't a possibility in the late 18th century. A respectable naturalist of the era no doubt felt he was on firm footing to pronounce that the creature was a hoax before he ever examined a specimen.

Of course, some scientists did, in fact, examine the specimens. They didn't find stitching or glue holding the parts together. Swiftly, the platypus was welcomed into the catalog of reality, and the textbooks describing what body parts a mammal might possess were, in fact, rewritten.

This is a useful insight into how science actually works. It blasts away a common idea espoused by some proponents of cryptozoology. There's a suspicion that evidence for a given cryptid exists, but this evidence is suppressed or sabotaged by scientists eager to preserve the current scientific consensus. Conspiracy-minded cryptozoologists might assume, since a giant, hairy, man-like beast in North America would complicate the overall theory of how, when, and where humans evolved, a scientist who runs genetic tests on a real Bigfoot bone is going to report that it's only a bear. A similar theory holds that the government has collected absolute proof of aliens. They have the crashed saucers, alien corpses, or are even hosting alien ambassadors, but they won't report it to the public because they don't want to create panic.

I can't claim that scientists are always paragons of veracity. There are numerous examples where scientists have adopted conclusions first and edited data to match these conclusions. For example, tobacco companies found numerous doctors willing to testify that there was no firm link between cigarettes and cancer.

Despite the fact that scientists possess human failings, the platypus demonstrates that, in science, truth will win out. If evidence contradicts current theories, it's an exciting opportunity to revise, refine, and retest those theories.

But, we're not here to determine the reality of honest scientists. Our subject for this chapter is the jackalope. Is it real, or not?

Physical evidence from modern times? Despite most stuffed jackalopes being taxidermy fakes, horned rabbits have been well documented. Random chance must occasional produce a rabbit with twin horns on its brow. Alas, no specimen of a virus induced horned rabbit is quite a match for the popular image of the modern jackalope, so this only earns a score of 16.

Fossil or other preserved evidence: I'm unaware of any rabbit fossils with horns, but, since we know horned rabbits can exist, it's not a stretch to assume that the drawing in the bestiary from 1575 was based on a real specimen, or reports of such a specimen. Since horned rabbits are diseased rabbits, they may not survive long in the wild. When a hunter encounters one, it's weird, and he'll talk about it. It hardly matters where the horn is placed. A horned rabbit is going to be a topic of discussion. The jackalope's appearances before exploding into greater fame in 1930 are likely to have a basis in this reality. I give this a score of 20.

Biological plausibility: Nothing beats the plausibility of living specimens housed at various zoos and research centers. While no example in captivity is known to have keratinous tumors positioned like antelope horns, nothing excludes them from growing in this configuration. This deserves a 20.

Habitat: There's a reason we say things "breed like rabbits." Rabbits and hares are well honed by evolution to exploit all manner of landscapes, from parched deserts with unbearable heat to frozen tundra where the food they feed upon is buried under deep snow a good part of the year. Introduced into environments where they never existed, like Australia, rabbits swiftly out-eat and outbreed native species. Since the virus that causes horn growth is also found on multiple continents, jackalopes are arguably the most far-ranging of all terrestrial cryptids. Another 20.

Concealment: The flip side of rapid breeding is that it typically evolves in species that don't live long. A hare with a horn blocking its vision or getting snagged as it tries to flee through brush isn't likely to become a scientific specimen. It's going to be dinner for a fox, hawk, or any number of predators. Some cryptids can be hidden in vast spaces, but this is a cryptid likely concealed by a brief lifespan. They are dead and eaten within weeks of their horns getting big enough to notice. Once more, a 20.

Strength of alternative explanations: The alternative explanation here is, at first glance, a strong one. As seen from the Herrick taxidermy creation, the overwhelming number of jackalopes on display are hoaxes. As a swift search of Amazon proves, jackalopes are a marketable concept. They are cute, kid-friendly cryptids. So, one alternative explanation of the jackalope is that the idea is kept alive simply because people like money. But, in the end, I'm giving the alternative explanations here a score of zero. They might explain the enduring popularity of the jackalope, but don't negate the verifiable existence of horned rabbits.

This brings the jackalope to a respectable score of 96. The primary wiggle room is that jackalopes aren't a distinct species. A purist can argue that if it's not a separate species, it's not a genuine cryptid. A man might be seven feet tall, extremely hairy, and go barefoot and naked in forests. We wouldn't call him a Sasquatch. My response to the naysayers is that they needn't be so picky. It's good to cultivate an openness to wonders. The proven existence of horned rabbits in the wild is enough for me to confidently declare the jackalope to be real.

The Reality Meter
Jackalopes

Modern Physical Evidence	16
Fossil Evidence	20
Biological Viability	20
Habitat Availability	20
Concealment	20
Alternative Explanations	0

In the case of the jackalope, a mountain of fake evidence can't conceal the underlying truth that rabbits sometimes, though rarely, grow horns.

MERMAID

Mermaids

When composing my list of cryptids to investigate, I was on the fence as to whether or not to include mermaids. They traditionally fall into the category of mythological creatures, and I don't think there are many cryptozoologists championing them as potentially real-world animals. I think their closest cultural comparison would be Santa Claus. People dress up in costumes to amuse children. Mermaid shows in Florida are a tourist draw, with relatively convincing costumes. Behind glass, through sun-dappled water, swimming mermaids can look quite real, just as the portly man in a red suit with a white beard can appear to be the real Santa Claus.

That said, people do still claim to spot mermaids. A quick search on YouTube turns up videos of mermaid sightings, though none are terribly convincing. As noted, mermaid costumes are available. Faking a video is no great challenge, especially a blurry one at a distance, or a poorly lit one in low-visibility water. Of course, as with the jackalope, the ease of faking evidence isn't evidence that all sightings are fake. Still, if mermaids are to be taken seriously, we need to use the same evidence standards demanded of other creatures. We need physical evidence, preferably a body.

It's this physical evidence that persuaded me to include mermaids in the book. Because, at one time, the supposed body of a mermaid actually achieved some degree of fame. The cryptid in question was known as the Fiji mermaid, because it was said to have been caught off the coast of Fiji. Badly preserved, it was still famously displayed by P. T. Barnum at Barnum's American Museum in New York in 1842. The mermaid was smaller than the typical creature of myth, only about three feet long. It was also somewhat less attractive than the

Fiji Mermaid

legend. Traditional mermaids are beautiful creatures who can lure men to their deaths. Barnum's mermaid looked more like a mummified monkey sewn onto a salmon.

Which, in fact, is what it was. Fisherman in Japan manufactured these taxidermy hybrids to sell to foreign traders. Whether the buyers thought they were purchasing genuine mermaids, or whether they were purchased as amusing novelties like the taxidermy jackalopes of today, I can't speculate. It's doubtful P. T. Barnum cared if his mermaid was a fake. People paid money to see it, and almost two centuries later people are still talking about it. Barnum definitely got his money's worth from the investment.

The Fiji mermaid can be dismissed as a fake, but it wasn't the origin of the mermaid legend. Artwork depicting creatures with human upper bodies and fishy lower bodies have been found in Babylonian artifacts over 3000 years old. Greek sirens were usually depicted as bird/human hybrids, but occasionally they appeared with fish bodies as well. Mermaids have a long, respectable lineage. Was there once a real creature underpinning the legends?

Fish and apes are very far apart on the evolutionary tree. It's difficult to think of an evolutionary pathway that would produce a hybrid creature with body parts that are a mix of fish and human. On the other hand, before the platypus was introduced to the larger world, people would have been skeptical that a creature could look like a blend of a duck and a beaver.

The platypus is semi-aquatic, as is the beaver. Numerous mammals had land-dwelling ancestors that adapted over time to life in the water. Some are so well adapted to water they live there full time, and have traded arms and legs for fins. In one of the many essays on whales contained in *Moby Dick,* Ishmael debates whether the whale is a fish or a mammal, and decides it must be a fish, because it shares so much in common with other fishes. If evolution can shape the dog-like ancestors of whales into something that looks like a fish, why can't it do the same to an ape?

Some people do live aquatic lives. The Bajau sea nomads live on the oceans near Indonesia and the Philippines. They live on houseboats and gather some of their food by diving. Over time, their bodies have adapted to allow them to dive deeper, hold their breath longer, and see beneath the water better than other humans. Evolution is shaping their bodies to their environment. With this proof that humans are capable of at least minor adaptations to aquatic life, is it such a stretch to imagine that some earlier sea-dwelling tribe of humans might have adapted even further, reshaping their bodies to permanently dwell at sea?

Unfortunately, the word "imagination" is doing a great deal of work in the previous sentence. If we can imagine how men might take on fishy features to live at sea, then it's just as likely that our ancestors might look at the sea and imagine what sort of men might live there.

Where does this leave mermaids on our reality meter?

Physical evidence from modern times: Unfortunately, beyond the obviously faked Fiji mermaid, I can find no claims of mermaid scales or bones intended to be taken seriously. For obvious reasons, we don't even have footprints. We must give this a score of 0.

Fossil or other preserved evidence: Nothing in the fossil record has yet been found even vaguely resembling a mermaid. But, depictions of aquatic humans appear in multiple cultures. There are many possible explanations for this. It's highly unlikely the depictions are based on real creatures, but not impossible. It's a slim chance, but I'll score this evidence as a 2.

Biological viability: Multiple mammals have made the transition from living on the land to living in the sea, so the idea that an ape-like creature could also make the journey is at least semi-plausible. None of the mammals who've made the journey have developed fish-like scales, but scales aren't unknown in the mammalian toolkit, as is evidenced by the pangolin. Most mammals that adapt to water wind up with short forelimbs, often little more than flippers, to become better swimmers. Still, given the usefulness of opposable thumbs, it's not completely unthinkable a creature could trade legs for fins but retain hands and arms. Though it involves a good deal of speculation, the fact that many mermaid traits are found in existing mammals means that the mermaid deserves a score of 5.

Habitat: There's a good reason whales, seals, otters and beavers adapt to live in aquatic environments. There's plenty of food to be had. The list of aquatic mammals is a long one. For this reason, we can score this a 20.

Concealment: Unfortunately, if we consider the sheer number of whale species that were hunted to the verge of extinction, we find proof that most aquatic mammals are unlikely to elude detection for long. Even the best adapted aquatic mammals breathe air, meaning they must spend a great deal of time near the surface. A man on a mast with a spyglass can spot them with relative ease. On the other hand, the ocean is a big place that still holds many mysteries, as we'll see when we discuss the kraken. Chances are, if mermaids were real, we'd have found them by now. Still, the vastness of their potential territory leads me to give this a score of 4.

Strength of alternative explanations: So far, I've talked about aquatic mammals without mentioning the most frequently cited alternate explanation for mermaids. One theory is that mermaids were simply manatees, perhaps seen by sailors from a distance, their heads draped in seaweed to resemble hair. As someone who's gone swimming with manatees and seen them up close, I feel like this alternative explanation is a little insulting to the sailors of olden times, not to mention the women of previous eras. A more likely alternate explanation goes back much further than the Babylonian depictions of merfolk. Cave paintings tens of thousands of years old depict humans, shamans perhaps, shapeshifting into other animals. The line between humans and animals hasn't always been drawn as strongly as it is in the contemporary world. Ancient art depicts men with the heads of animals, or animals with the faces of men, with the Minotaur and the Sphinx among the many famous examples. Merfolk were likely just one of many possible man/animal blends imagined in these ancient cultures. Thanks to art and storytelling, echoes of these beliefs endured into relatively modern times. I think the likelihood that the mermaid is an echo of earlier pagan blurring of the lines between man and beast is quite high, at least worth 18 points, which must be subtracted from our total. The only reason I'm not scoring this a twenty is that, until someone invents a time machine, we can't actually ask the artists of the ancient world what truly inspired them.

This produces a score of 13. If a convincing specimen of mermaid turned up tomorrow, it would require drawing new branches on the evolutionary tree. Still, I imagine we'd get used to their existence quickly enough. Mermaids have been part of human culture for thousands of years and show no decline in popularity. Will newcomers like the Chupacabra or the Lizard Man of Scape Ore Swamp prove to be as enduring?

The Reality Meter
Mermaids

Modern Physical Evidence	0
Fossil/Preserved Evidence	2
Biological Viability	5
Habitat	20
Concealment	4
Alternative Explanations	-18

LIZARD MAN

Lizard Man

ermaids have been with us seemingly forever, but the Lizard Man of Scape Ore Swamp near Bishopville, South Carolina, is a cryptid of recent vintage. The first stories only date from 1988, and the initial report didn't even feature a lizard man sighting. Instead, the sheriff's department was called out to investigate a damaged car. The car was scratched up and punctured with what looked like tooth marks. Muddy footprints were found at the scene.

When news of the damaged car was reported locally, a 17 year old named Christopher Davis filed a report that his car, too, had been damaged, and that he'd seen the culprit—a seven foot tall creature with three fingers, red eyes, and scaly skin like a lizard that jumped onto his car near Scape Ore Swamp.

This story spread rapidly. More and more people came forward with claims that they, too, had seen the lizard man. Media reports drew tourists to the area, and the locals were happy for the attention. Lizard man t-shirts appeared in local stores. Eventually, the creature was "investigated" on TV shows like *Paranormal Files: Fact or Faked,* and *Destination: Truth*, further increasing the Lizard Man's notoriety. In recent years, the town has hosted an annual festival called the Lizard Man Stomp.

And… that's pretty much it. There's not much else to relay. But I've chosen to include Lizard Man in the book for two reasons. First, I kayak quite a bit in South Carolina, and have ventured into my share of swampy backwaters full of alligators. There's beauty in these places, but also an air of sinister mystery that leads anyone with an active imagination to wonder what monsters might be lurking beneath the murky waters.

An alligator the author has met in South Carolina says, "Howdy!"

Second, there's an important insight here for anyone who studies cryptids. The entire legend of the Lizard Man grows out of a damaged car and a single unverifiable sighting. Subsequent sighting echo the original story. Physical evidence was limited to scratch marks and muddy prints. From such thin evidence, the story still gets coverage in state and local papers thirty years later. Cable channels have devoted time and resources to the investigation and found or manufactured enough content to fill episodes. Once a cryptid gains even a little media coverage, it doesn't take much to keep feeding the legend.

But is there any reason to think it might be more than just a legend?

Physical evidence from modern times: The only noteworthy physical evidence was the first scratched up car. But, considering that there are plenty of known animals capable of scratching up a car—including mischievous humans—this is hardly conclusive. I will generously score this a 1.

Fossil or other preserved evidence: Before 1988, there's no earlier reports of a Lizard Man in South Carolina that I can locate. The fossil record does have quite a few bipedal reptiles in it, mainly dinosaurs. It's a real stretch, but since we know from fossils that some reptiles can walk on two feet, I'll give this a score of 2.

Biological viability: I'll once more invoke the platypus as a caution against saying it's impossible that there could be an unknown animal waiting to be discovered that blends the bodily features of both a man and a reptile. Some mammals have scales. Some lay eggs. Still, if a seven foot tall bipedal lizard was discovered in South Carolina, it would require some serious work to fit it into the evolutionary tree. We'll stretch to give this this a 4.

Habitat: I've driven through Bishopville. Scape Ore Swamp simply isn't very big. You can pull it up on Google maps and see for yourself. If there really is a Lizard Man, there must also be lizard women and children. A single swamp couldn't possibly support a breeding population. That said, South Carolina is home to some sizable alligators. Scape Ore Swamp drains into the same basin that houses Lake Marion and Lake Moultrie, where alligators grow to be over 13 feet long. It requires extending Lizard Man's range quite a bit, but if SC can support alligators twice the length of a grown man, it could, in theory, support a man-sized, bipedal lizard. However, since the largest reptiles wouldn't make their home in a body of water as small as Scape Ore Swamp, I feel like the best I can score this is a 10.

Concealment: Lord no. If you live in a city, a small town like Bishopville might seem like it's literally the middle of nowhere, surrounded by miles of barren, empty swamps. In reality, I-26 passes just south of Bishopville, and Highway 15, a major state highway, cuts through the area. What wilderness remains is well trodden by hunters, and any water deep enough to conceal a lizard man is deep enough that someone is taking a fishing boat or kayak there. This isn't inhospitable Siberian wilderness where no one ventures. If this is a reptile, South Carolina is chilly enough it would need to hang out on open banks to sun itself, just like the alligators. My firsthand experience with the terrain leads me to give this a score of 0.

Strength of alternative explanations: There really are few alternative theories to explain the Lizard Man, mainly because there's so little to explain. However, tossing out the obvious alternatives that this is simply a hoax or a tourist trap, we can always appeal to our old friend pareidolia. The swamps have strange shadows. You can see all manners of things that aren't really there. We'll be subtracting 10 points because of this.

Thus, the Lizard Man of Scape Ore Swamp comes in on the reality meter with a paltry 7 points. It's dubious that anyone truly thinks such a creature exists. But, if my schedule permits it, I'd love to make it to the Lizard Man Stomp one day!

The Reality Meter
Lizard Man

Modern Physical Evidence	1
Fossil/Preserved Evidence	2
Biological Viability	4
Habitat	10
Concealment	0
Alternative Explanations	-10

THE LOCH NESS MONSTER

The Loch Ness Monster

If Bigfoot has a serious challenger to the title of most famous cryptid, it has to be the Loch Ness Monster, AKA, Nessie. With sightings going back centuries, a plethora of eye-witnesses, and countless out of focus videos and photos, Nessie is the subject of some of the most serious and well-funded cryptid searches ever undertaken. Sonar scans have returned images showing objects 30 feet long moving beneath the water. There has, alas, never been an actual specimen collected. Worse still, DNA analysis of water samples from Loch Ness have identified nothing anomalous. The species known to inhabit the lake were the only species found in the water.

Could it be there was a Nessie until relatively recent times, but she was no longer in the lake by the time the DNA samples were taken? Perhaps. But Nessie is hardly the only lake monster. Numerous bodies of water all around the globe have a claim to large, unknown aquatic beasts. If one of them has passed away, there's still hope for finding a related specimen elsewhere.

But the widespread distribution of lake monsters creates a new problem. How can creatures thirty feet long migrate all over the world? One solution is that huge fish can still lay tiny eggs. Adult aquatic giants might start life smaller than a pea. The life cycle of creatures like the eel are proven to involve massive migrations across thousands of miles, moving between freshwater and saltwater environments. Big lake monsters could theoretically colonize fresh bodies of water while still in larval or juvenile stages when they are little bigger than minnows, and we simply don't notice them.

If fans of Nessie and her kin were content to believe that lake monsters were simply large fish, there would be little debate as to whether they might exist. Freshwater fish can grow to impressive sizes. The current world record for a catfish is nine feet long. In the not so distant past, before so many of the world's rivers and lakes were degraded by pollution, some catfish might have overshadowed today's specimens. Modern sturgeon have been found over 20 feet long and weighing up to 3000 pounds. It's not a huge leap of faith to think that specimens longer than 20 feet might have once been plentiful, or that a few especially lucky individuals might still lurk in deep waters. Alas, with each passing year of environmental stress, the odds that truly massive freshwater fish can survive are dwindling.

Of course, most lake monster fans aren't satisfied with the plausible explanation that their favorite cryptids might be nothing more exotic than a sturgeon or catfish. Since the famous "Surgeon's Photo" of 1930 showed a shape like a long, curved neck rising from the water, Nessie and her fellow lake monsters have been identified in the popular mind as some remnant species of plesiosaur.

Identifying the Loch Ness Monster as anything other than a fish presents numerous challenges to plausibility. It the same difficulty that arises in concealing a mermaid. A reptile can remain underwater longer than a mammal, but they still must surface to breathe. On the smooth, calm waters of a lake, with a good zoom lens camera, we'd be able to photograph a breeding population of such creatures several times each day. Aquatic reptiles face a further challenge as a Loch Ness Monster candidate. They're cold-blooded. The large aquatic reptiles we do know of, like crocodiles, not only surface to breathe, they crawl out of water to bask in the heat of the sun. Large reptiles like crocodiles or boas are also limited by climate. Cold water provides a cutoff for how far north an aquatic reptile could plausibly live, and Scotland, alas, is quite far above that line.

While I'm reluctant to compare a respectable cryptid like Nessie to a newcomer like Lizard Man, there are a few noteworthy similarities to consider. In both cases, the fame of the monster makes eyewitness testimony difficult to judge. Despite declaring the Lizard Man to be definitively unreal, I confess that if I were to be driving along the edge of Scape Ore Swamp, I'd be watching the trees, looking for anything unusual. Given the fame of the Loch Ness Monster, even the most hardened skeptics can't help but take note of every ripple in the water when they stand upon the lake's shore. If people are looking for something odd, all sorts of mundane objects can turn into mystery beasts.

Second, while Lizard Man has given birth to a local festival that draws visitors for one day each year, Loch Ness draws tourists year round, many of them hoping for a glimpse of the famous beast. I can't help but think that if the lake was completely drained and the muck thoroughly searched, it wouldn't matter whether or not any monster was found. The shops of Loch Ness would still sell shirts and caps emblazoned with the creature's image. The second the lake was refilled, the sightings would resume. There's a massive economic incentive to pretend there's something fantastic and wondrous lurking in the waters, and no reward at all for declaring the lake 100% monster free. If the monster hunts of today fail to produce a specimen, don't worry. There's always tomorrow, and tomorrow, and tomorrow. Your great, great-grandchildren will grow up hearing the legend of Nessie, and newspapers will report on the latest expedition hunting the beast. Assuming there are still newspapers. Given current trends, newspapers will vanish from the world long before Nessie.

If I sound cynical about Nessie's existence, I assure you that doesn't keep me from being a fan. The fact that the lake has been searched with sophisticated sonar equipment, explored by cutting edge submersibles, and sampled for DNA shows that even a mythical monster can motivate people to study the world more closely. If Nessie and her submerged kin inspire new generations of scientists, they're performing a great service.

Let's set aside both my cynicism and my admiration. What does the reality meter say? In a surprise twist, we'll be doing the analysis twice! First, let's consider the possibility that Nessie and her fellow lake monsters are related to plesiosaurs:

Modern physical evidence: Radar traces of something big on the bottom aren't enough. Eyewitness accounts and blurry photos don't count. I'm afraid this is a 0.

Fossil and other preserved evidence: Obviously, we have fossils of plesiosaurs, but they are very far removed from the modern era. 200 million years is quite a gap in the fossil record if some subspecies did survive. At best, this earns 10 points.

Biological viability: While we can't dispute that creatures like plesiosaurs can exist on Earth, they thrived in a much warmer era. The cold-blooded nature of reptiles means that, biologically, they are a poor fit for Northern lakes. Let's call this a 6.

Habitat: While cold water is a poor fit for reptiles, cold, deep lakes often have plentiful fish. It's plausible one or two big predators could share a body of water the size of Loch Ness without risking starvation. This abundant food source lets us score this 8 points.

Concealment: It's plausible that a big reptile might have eluded detection before modern times. When photos were still taken on film, bad luck and bad lighting might have thwarted getting a good picture. But, since reptiles need to breathe, hiding a big one in the modern world from all the drones, mounted cams, sonar scans, and DNA tests available pushes the odds of eluding detection past the point of believability. This earns a 0.

Alternative explanations: It could be a big fish! Or it could be a fun local legend that sustains itself with fresh sightings whether or not anything is actually seen. I'm going to give these collective theories 18 points. Subtracting those out brings the plesiosaur interpretation down to a 6, placing it in Lizard Man territory. Entertaining, but not to be taken seriously.

Now let's put a big fish on the reality scale:

Modern physical evidence: While encounters with them are increasingly rare, numerous freshwater fish have been caught with lengths longer than 10 feet. Ten feet might seem small for a monster, but my heart would race if a fish longer than my kayak surfaced next to me. Distances and sizes on open water can be deceiving, so the 30 foot lengths attributed to some lake monsters shouldn't be taken as gospel. Casting a broad net, the proven existence of monstrous fish scores a 20.

Fossil and historical evidence: Fish are one of the most abundant and successful life forms ever to evolve on the planet. They adapt to habitats where no reptile or mammal can hope to survive. In fact, it's not a stretch to say that reptiles and mammals are both very advanced versions of fish adapted to survive above water. Unlike plesiosaurs, fish fossils don't vanish from the fossil record 200 million years ago. They're still here, and are likely to carry on even if humanity faces extinction. This category earns a solid 20 points.

Biological viability: At the risk of redundancy, large freshwater fish have been collected in recent decades. An easy 20 points.

Habitat: Northern deep water lakes contain plenty of small fish, which happens to be the favorite meal of big fish. Unlike the reptilian theory of lake monsters, cold is no barrier to a fish's range. Finally, for spreading from lake to lake, the eggs of even the biggest fish are rather small, making birds occasionally transport them to isolated bodies of water. Another 20.

Concealment: Not much better than the plesiosaur, alas. I won't score this as a 0, since fish have the concealment advantage of never needing to come to the surface. Still, between sonar, DNA testing, submersible drones, and modern photography, the odds that there's an unknown species of big fish still hiding in a lake somewhere is low. Of course, we don't need an unknown species; known species like the sturgeon can play the role of lake monster. But, given how skilled humans are at finding fish in even the deepest of inland waters, I can only give this a score of 2.

Alternative explanations. It could be a plesiosaur! Wait, we've covered this. I understand the appeal of wanting some wondrous and unknown creature to explain the strange bulges and wakes we see in remote lakes. But, I'm unwilling to rewrite the fossil record to champion an exotic creature when more mundane creatures, of proven reality, can easily fill the role. The even more mundane possibility exists that people are seeing logs, or waves, or simply making things up. But, even if some of this does happen, that doesn't negate the known existence of monstrously large fish. I'll only give the alternatives 5 points to deduct from our total.

This brings the "big fish" version of lake monsters to a respectable 77. Nessie herself might no longer be real, but the possibility that there are creatures larger than boats lurking beneath the waters of a large northern lake isn't just plausible, but likely.

The Reality Meter

Lake Monsters — Plesiosaur

Modern Physical Evidence	0
Fossil/Preserved Evidence	10
Biological Viability	6
Habitat	8
Concealment	0
Alternative Explanations	-18

The Reality Meter

Lake Monsters — Big Fish Version

Modern Physical Evidence	20
Fossil/Preserved Evidence	20
Biological Viability	20
Habitat	20
Concealment	2
Alternative Explanations	-5

MOTHMAN

Mothman

I've dreaded writing this entry. I *like* Mothman. I've driven hundreds of miles to hunt for him! More accurately, I went hunting for his statue. I found it quite easily, right off the main street Point Pleasant, adjacent to the Mothman Museum.

Like the Lizard Man, Mothman is a fairly recent arrival on the cryptid scene. Before 1966, there were no reports of a man-sized creature with wings and large, glowing eyes haunting the mountains of West Virginia. Also like Lizard Man, an initial sighting was swiftly followed by stories of more sightings. These reports usually didn't mention seeing a large, winged man, but instead called the creature a bird.

In the pre-internet era, it's possible these initial stories might have faded into memory before reaching the broader public. But, in 1967, the Silver Bridge in Point Pleasant collapsed, killing 46 people. A disaster of this magnitude brought national media to Point Pleasant. Reporters heard the local Mothman legend, and the two stories have since been tangled up in the public's imagination, with the Mothman sometimes portrayed as a supernatural entity warning of doom, a modern angel of death.

Angels, alas, are outside the scope of this book. We can't test the DNA or dissect the bodies of spiritual creatures, so they must, for now, reside outside the realm of science. We can set aside exotic explanations such as the possibility that Mothman is an alien, a time-traveler, or a one-off mutant escaped from a government lab. Any of these theories is based on unproven assumptions that can neither be confirmed or refuted. Instead, we'll treat Mothman like we treat other cryptids, as a potentially real creature, part of a larger population that eats and breeds and only shows up near the site of bridge collapses out of pure coincidence.

The Mothman depicted by the statue, not to mention portrayed on bumper stickers and car mugs, doesn't make a lot of sense, scientifically. A platypus, with its mix of duck and beaver parts, makes evolutionary sense when you realize that ducks and beavers actually adapt to live in similar environments. Duck bills are handy for finding food in murky water, webbed feet are handy for paddling around. Convergent evolution leads numerous unrelated species to share traits. Bats and birds get to flight through different evolutionary pathways, but in the end they both have wings.

Humans haven't had much evolutionary pressure to grow wings, no matter how much we might fancy them. We've invested our evolutionary capital in large brains, agile hands, and upright posture. So far, the investment seems to have paid off.

Winged insects are also getting good returns on their evolutionary investment. Their wings help them elude predation and carry them to the resources they need for mating.

At some point in the last 600 million years, humans and moths must have had a common ancestor, and we still share a surprising amount of genes with insects. But, in the end, human body parts make sense on a human body, and insect body parts only make sense on insects. Insect wings, if scaled up to hang glider size, would never get a human body into the air. Big brains in an insect would be a tragic waste of metabolic resources. Creatures like mayflies use the minimum possible nerve pathways to mate and lay eggs during their few hours of life as a flying creature. Spare brain cells that let them contemplate their mortality would only slow them down.

The platypus is proof that odd things exist in nature, but it's not proof that you can just jam together body parts from different creatures and have a functioning animal. A manlike creature with moth wings would still be too heavy to fly. An insect scaled up to human size would collapse under its own weight. The largest insect nature has ever produced, a prehistoric dragonfly, had a wingspan of almost 30 inches. This is impressive, but it's not something likely to be mistaken for a man.

We may just be getting tripped up by assuming that Mothman has the body parts of either a moth or a man. There are numerous large winged animals in the world. Could the initial reports of a large bird be the plain answer to the Mothman mystery?

At first glance, there aren't any candidate birds in the area that would grow big enough to be mistaken for a man. However, migratory birds can be blown off course. Could Mothman have simply been a large bird not native to the area? It seems unlikely that unlucky winds could have blown a condor all the way in from California. Still, unlikely isn't the same as being impossible. To someone in West Virginia who's never encountered a condor, having a bird with a wingspan wider than your car pass by your windshield must make quite an impression.

Yet, the idea that the Mothman legend springs from nothing more than some large, out of place bird feels off. It feels too easy. Big birds are seen all over the place without spawning enduring legends like Mothman. Certainly, there's something stranger going on here.

Maybe. But, there's something perverse about human nature that lets us reject unlikely scenarios only to embrace impossible ones.

I'm going to detour from Mothman to tell a true life ghost story. I used to live near Hopewell, Virginia. A major Civil War battle was fought nearby, and many historic houses from the era still stand in the historic district. I had a friend who lived in one of these houses, very near a park commemorating the Civil War. The house had been commandeered for use for a field hospital during the Civil War. My friend had a historic photograph of Robert E. Lee and other officers standing on his porch.

Old houses, especially ones near battlefields, often come with ghost stories associated with them. My friend swore his house was haunted, and that he'd seen the ghost. He'd been alone in the house watching television one day when he'd gotten a strange feeling. Looking around, he saw a woman dressed like a nurse from the

Civil War era inside his house, walking up the steps to the second floor. I asked if he'd asked the woman what she was doing in his house. He said he didn't ask questions, and had instead gone into another room, waited a few minutes, then worked up the courage to go up the stairs. No one was there. He'd plainly seen a ghost.

I asked why he thought the woman was a ghost. He said she'd looked exactly like some of the nurses in the historic photograph of his house. I conceded this was completely possible. Civil war reenactors wearing period garb were hardly rare in that area. Many gave tours of the local battlefields. Some, conveniently, gave ghost tours of the surrounding area. A modern woman dressed in Civil War garb was certainly something he'd seen before. He lived in a historic house adjacent to a public destination that attracted Civil War buffs. Wasn't it more likely that he'd seen a living woman than a ghost? Maybe she'd entered the house through an unlocked door accidentally believing it to be open to the public? Maybe she'd come into the house a little ways, realized this was someone's home, and quietly turned around and left without saying a word?

My friend thought this was absurd. I was really grasping at straws constructing such an unlikely scenario. I admit, it requires a lot of assumptions. There's no way to prove it happened as I imagined it happened. We were both offering speculation. I was speculating he'd seen something that indisputably exists, a woman in costume. He speculated that dead spirits can manifest themselves long enough to walk upstairs before vanishing once more into the ether.

I'm often reminded of this conversation when I'm contemplating cryptids. Someone will offer an unlikely explanation that can never actually be proven, like proposing that a very large bird wandered thousands of miles outside its habitat and spooked some locals. People's skepticism justifiably kicks in. There's no proof that actually happened! But if someone proposes that there's a creature the size of a man that also has mothwings and glowing eyes and is possibly a harbinger of doom… sure. That seems like a real possibility.

Is it pure entertainment value that makes the second scenario so much more appealing? Is it some hunger to encounter something that doesn't comply with the mundane boundaries of ordinary reality? Perhaps our skeptical facilities function fine up to a certain level of probability, but break down beyond that? If the last possibility is correct, it could explain a great deal of human history.

Let's leave behind speculation, and run Mothman through the reality meter!

Physical evidence: For such a well-known cryptid, it's weird we don't even have a few ambiguous footprints in damp mud. 0.

Fossil and historical evidence: Nothing in the fossil record shows up as a blend of ape and insect. While winged men do have some pedigree in ancient art and cave sculptures, not to mention countless depictions in religious iconography, there doesn't seem to be any specific tradition of winged men in West Virginia predating 1966. This also earns a 0.

Biological viability: Mothman's an anatomical mess. He's much too big to be flitting around on insect wings. He doesn't work as an evolved bug, and he doesn't work as a mutant man, leaving him stranded zoologically. If a genuine specimen turned up, every working biologist would be forced to wonder if everything they knew about natural selection was wrong. Another 0

Habitat: Not knowing what Mothman eats, it's difficult to judge. The area around Point Pleasant hosts abundant wildlife. Without more information on the creature's biology, I can't give this a maximum score, but whether it's a carnivore, herbivore, or omnivore, I think it could find at least something to nibble on. I'll be generous and give this a 5.

Concealment: West Virginia is pretty rugged. But, given his size, Mothman can't be flying below the forest canopy. If he travels above the trees, even at night, it's difficult to see how he's not spotted more frequently. If he spends most of his time on foot, concealment is even less likely. West Virginia has a lot of hunters, and the woods are no doubt bristling with trail cams. My last gasp speculation is that, if Mothman is of insect origins, perhaps his kin only hatch and fly around for a few hours of mating, then spend years below ground as larvae, similar to cicadas. I know I'm piling speculation on speculation with this larval theory, but I want to give Mothman at least some minor hope of reality. I feel like I'm stretching, but I'll score this a 5.

Alternative explanations: There's still a real possibility people simply saw a large, unfamiliar bird. In 1966 you couldn't snap a photo with your cell phone and google what the heck you just saw. Birds exist. Large birds sometimes turn up where you don't expect them, since winds can carry them far beyond their native range. The major flaw with this alternative is that, if it was a bird, it's odd that anyone would have mistaken it for anything else. When I'm out kayaking or hiking, I often see birds I can't identify, but I have no doubt that I'm looking at a bird. For this reason, though I'm fond of the bird theory, I think I can only give it 8 points to deduct from the total.

Sorry, Mothman. With the reality meter returning a score of only 2 after all is said and done, it's my sad, heartbreaking duty to inform you that you don't exist and have never existed. But, I'm still going to wear my Mothman t-shirt.

The Reality Meter
Mothman

Modern Physical Evidence	0
Fossil/Preserved Evidence	0
Biological Viability	0
Habitat Strength	5
Concealment	5
Alternative Explanations	-8

THUNDERBIRD

Thunderbirds

The possibility that Mothman sightings might have been inspired by nothing more than a large bird opens the door to another category of cryptids. What are we to make of legends of enormous birds much larger than any known species, such as the fabled thunderbird?

I want to tread carefully here. For certain indigenous tribes of North America, the thunderbird held spiritual, even divine significance. It could be seen as disrespectful to treat thunderbirds as merely unknown animals and judge their existence through a purely scientific lens. It's a bit like categorizing angels or demons as cryptids. Many people believe in such beings as a matter of faith, and may even find such constructs useful for navigation through the world.

I'm tackling the thunderbird as a physical creature, the product of evolution, for the simple reason that the claim that the wilderness is populated by enormous birds isn't unique to North America tribes. Giant birds are found in folk legends from many cultures, with perhaps the most famous being the roc of the *Arabian Nights*. The roc also makes an appearance in *The Travels of Marco Polo,* where the creature was said to be:

> *"... for all the world like an eagle, but one indeed of enormous size; so big in fact that its quills were twelve paces long and thick in proportion. And it is so strong that it will seize an elephant in its talons and carry him high into the air and drop him so that he is smashed to pieces; having so killed him, the bird swoops down on him and eats him at leisure."*

Marco Polo recounts the tale of the roc as factual. But is there any evidence that truly massive flying birds lived into modern times? A great deal depends on how big you feel a bird must be in order for it to impress you. Some albatrosses have wingspans nearly 12 feet long. The California condor has a wingspan a little shy of 10 feet across. Even small birds can cast sizable shadows. If the shadow of a condor passed over you, you might be forgiven for overestimating its size.

Of course, condors are scavengers and albatrosses eat fish and squid. Neither prey on people, let alone elephants. Do we have any other candidates that can stir up a bit more concern?

I present to you the *Argentavis magnificens,* the largest bird ever discovered in the fossil record. It had a wingspan a little over twenty feet, making it twice the size of a California condor. That's also a length close to four times the height of prehistoric humans, which would make a close encounter with one rather intimidating. It's not an elephant-dropper, but we're reaching the size of small airplanes. If one were discovered living in the wild, I think quite a few people would be

Argentavis magnificens compared with a man of average height.

content to label it as a potential thunderbird or roc. The one tiny barrier to *Argentavis* being our cryptid? It went extinct at least 10,000 years ago. It also lived in South America, mostly in a period of time before humans reached the most remote portions of that continent. Even if we allowed for some overlap in time between humans and *Argentavis,* it's difficult to see how sightings of the bird in South America could be the origin of the roc legends of Asia. In the absence of any more recent physical evidence, rocs and thunderbirds seem to belong to the spiritual realms of angels rather than the mundane world of condors, albatross, and eagles.

The reality meter:

Modern physical evidence: Pretty much nothing. 0.

Fossil or historical evidence: If we can find *Argentavis* remains, it's certainly plausible earlier cultures could have stumbled across similar bones and fossils, inspiring the legends. This earns a 10.

Biological viability: The largest pterosaur fossils ever found were of *Quezocoatalus*, which possessed a 40 foot wingspan. This gives us proof that physics and biology permit would allow for the existence of some truly monstrous flying birds. Alas, our elephant snatching roc is well beyond the upper limits of the possible. I'll again give this a middle score of 10.

Habitat: Sadly, the modern world is no longer a place where large birds thrive. Birds are sensitive to even minor changes in their environment, and humans are capable of altering landscapes across entire continents. The largest surviving bird, the California condor, is hardly thriving. So, even though large birds may have soared through the skies in prehistoric eras, today's world seems like a poor fit. I can only give this a score of 2.

Concealment: A truly large bird could be seen from miles away while in flight. There simply aren't that many regions of the world that humans don't occupy or at least traverse. Even the most barren deserts and densest jungles are flown across by airplanes. It's true some birds fly at night, when they'd be a little tougher to spot, but that doesn't keep us from locating and filming owls in flight. Finally, concealing a bird of almost any size is made trickier by the fact that bird-watching is an avid hobby for millions of people who enthusiastically seek out and record different species of birds. There are more bird-watchers than Bigfoot hunters, and they have a much better track record of finding their subjects. This earns a 0.

Alternative explanations: The strongest alternative explanation of the North American thunderbird is simply that it's a spiritual tradition that lies outside the realm of scientific study. The roc is only a traveler's tale. But, since we weren't there when these legends originated, it's tough to pinpoint how much was pure invention, and how much of the myth, if any, took inspiration from large birds of previous eras. I'll give this another middle score, and deduct 10.

Add it all up, and thunderbirds and rocs score a 12. It's vaguely plausible that some unknown species with a wingspan larger than that of the California condor survived until relatively recent times. If remains of such a creature are found mummified in desert sands, it will be an exciting discovery, but won't require tossing out all we know about the natural history of birds.

The Reality Meter
Thunderbirds

Modern Physical Evidence	0
Fossil/Preserved Evidence	10
Biological Viability	10
Habitat Strength	2
Concealment	0
Alternative Explanations	-10

DRAGON

Dragons

Like mermaids, dragons straddle the line between pure mythology and cryptozoology. There aren't many people still searching for dragons in the wild. But, they appear so often in old art and literature that it's plain that the beasts seemed real to a significant number of our predecessors. The *King James Bible* mentions dragons over thirty times. For Christians in medieval Europe this must have seemed like the final word on the creature's reality.

It's apparent from older art that people weren't quite certain what, exactly, a dragon might be. Was it only a big snake? Dragons and serpents seemed to be used interchangeably at times in the Bible. Snakes can reach monstrous size, with the largest being a 33 foot long anaconda discovered in Brazil. The problem with identifying a snake as a dragon is that our ancestors in all but the coldest climates would have been familiar with snakes, and wouldn't have needed a completely different category of animal to describe them. Dragons were also said to be poisonous. In addition to numerous venomous snakes, there's also a species of giant monitor lizard with a venomous bite–the Komodo dragon. Yet, the word dragon was being used centuries before the Komodo dragon came to the attention of westerners. It could hardly have been the inspiration for the dragon of myth.

Then there's the notion that dragons could fly. The book of Isaiah references a flying, fiery serpent. Though not specifically named as a dragon, the notion that dragons were associated with fire and also that they could fly might find its origin here. Wherever these draconic attributes originated, a red dragon came to stand as a symbolic representation for the nation of Wales. This Welsh dragon seems to have set a standard of what Europeans expected a dragon would look like.

The Welsh dragon

All the classic elements are here. It's scaly with a long neck and long, coiling tail, which makes it part serpent. It has wings and red coloration, making it fiery and flying. For good measure, it's given a somewhat leonine body, including a full complement of long, sharp claws ready to tear apart its prey. Whether the first person to draw this dragon thought he was depicting a real creature, or was simply trying to draw the most intimidating beast he could think of, we'll never know. But, for the purpose of this chapter, this is the dragon archetype we'll focus on. (I'm sidestepping Chinese dragon myths or dragons from other cultures, since all may have different origins.)

Was there ever any reality to the classical, four legged, winged, fire-breathing dragon? If not, just how many classical dragon traits can a creature give up before it's reduced to a mere snake or lizard?

The first strike against the classical dragon is that it has six limbs—four legs and two wings. The evolutionary toolkit available to terrestrial vertebrates makes this configuration almost impossible. The first fish to scramble onto the shore had four limbs. Since then vertebrates have been able to successfully shed limbs (as seen in snakes), but no vertebrate has ever added limbs. Most vertebrates are modeled after that first fish, supporting themselves on four-limbs that come into contact with the ground. Vertebrates that fly, like birds and bats, have their forelegs sculpted into wings by evolution. Humans and monkeys have their front legs transformed into arms, and feet converted into hands. If you strip away the flesh from chickens, fruit bats, whales, dogs, and chimps, the bones tell a common story.

Most creatures work with the same basic skeletal framework. Different lengths of bones adapt the limbs to different uses. You find echoes of the chimp's fingers in the bony struts of a batwing. We think of dogs having backward knees, but what we think of as their knee is the equivalent of our ankle. The bones beneath that backward bend in a dog's legs serve as the dog's lower leg, but the same bones in humans form our feet. The repurposing and adaptation of skeletal structures is one of the most important lines of evidence for evolution.

Some medieval artists seemed to be aware of this, and drew dragons with two legs and two wings. This puts us back into the realm of the plausible. More than plausible, in fact. We have fossil proof that reptiles resembling these beasts existed long ago. Before there were birds, there was the archaeopteryx, a dinosaur relative with long, toothy jaws, a tail, and feathered wings. If you demand scales on your dragons instead of feathers, please note that feathers are merely scales modified by evolution. (The same is true of hair.) If an archaeopteryx had somehow survived long enough to overlap with modern humans, it could have provided a convincing template for dragons as a toothy, flying reptile. Alas, they died out 150 million years before humans ever came onto the scene. Still, the fact that they existed is proof that a relatively convincing dragon could have a spot in the Earth's current catalog of beasts.

Of course, no one is claiming that the archaeopteryx could breathe fire. But, how essential is fire-breathing to the draconic formula? Would another form of ranged attack originating from the mouth suffice? Cobras spit venom at the eyes of attackers. Since biological science gives a stamp of approval to venom spitting as a proven mode of attack, we can assign it to some undiscovered species of archaeopteryx and have ourselves a creature that I think would pass as a very convincing dragon.

Now that I've convinced myself that dragons are biologically plausible, do I think they've ever existed? No. So what explains the myth? Why are dragons so enduring? Why can we look at an unrealistic creature like a Welsh dragon and instantly recognize it as a fearful predator?

I once more invoke evolution. Prey species have instinctive fears of predator species. Humans live in a world where we no longer worry about predation, but we evolved from smaller, tree-dwelling ancestors that had multiple creatures hunting them. Pythons eat monkeys. Our ancestors had good reason to be on the lookout for big snakes. While people can learn to love snakes, many people also demonstrate deep-seated, irrational fear. Monkeys also get snatched up by eagles. An innate fear of birds larger than us might contribute to the myth of thunderbirds and rocs. Finally, monkeys are preyed upon by big cats like lions and leopards. Suppose you were hiking in remote mountains and realized that a panther was coming down the trail toward you. You might rationally think that the panther will be more scared of you than you are of it. But, underneath your rationality, you'd still have a rush of adrenaline from the presence of a wild predator.

Our ancestors left us with hardwired fears of lions, eagles, and snakes. Blend the traits of these creatures into a single being, and you get a dragon. No wonder they seem so intimidating.

Let's go to the reality meter:

Modern physical evidence: Snakes, crocodiles and alligators kill people every year. They might not be winged lizards dive bombing us while blasting us with fire, but the world is good at producing dangerous predators covered with scales. I think this deserves at least 5 points.

Fossil and historical evidence: The archaeopteryx makes for a passable dragon. Fossils also help us establish the evolutionary gateways creatures must pass through, ruling out the hexapod dragon of the Welsh flag. It's a mixed bag, so I'll give this a 10.

Biological plausibility: While I don't see how you assemble a fire-breathing hexapod dragon from evolution's parts box, a winged reptile with a breath weapon isn't completely out of the question. I'll give this another 5.

Habitat: Giant lizards had a good run! Dinosaurs and their kin dominated land, sea, and sky for 165 million years. If not for an asteroid, they might still be at the top of the food chain. Still, many large predators have come under intense pressure from humans, both because we encroach on their territory and because we actively seek out and kill predators that are a threat to us or our livestock. I feel like this deserves another middle score of 10.

Concealment: New species are discovered all the time, but discoveries of really big creatures are rare. A dragon would be an apex predator, leaving scat and bones all around. I just don't see how it avoids leaving behind evidence any hunter with a dog can easily find. I'll give this a 0.

Alternative explanations: I like my "evolutionary composite fear beast" theory of dragons. It would be interesting to show monkeys images of dragons to see if they react fearfully. But, I don't have the resources to conduct this experiment, and even if I did I'm too kind-hearted to go around stressing out monkeys to satisfy my curiosity. All I have is speculation. The most likely alternate explanation is that the dragon was always a creature of myth, no more real than a minotaur. I'll subtract 10 points.

We arrive at a reality score of 20. Large, dangerous, scaly predators that share traits with dragons have existed in prehistoric times, and a few intimidating reptiles still share the planet with modern people. Despite this, we can say with strong confidence that dragons don't exist today, and that nothing we'd indisputably label as a dragon has existed alongside humans. My only hope of one day meeting a real dragon lies in genetic engineering, but that's a subject for a completely different book.

The Reality Meter
Dragons

Modern Physical Evidence	5
Fossil/Preserved Evidence	10
Biological Viability	5
Habitat Strength	10
Concealment	0
Alternative Explanations	-10

THYLACINE

Thylacines

In the previous chapter, I made the claim that large predators come under intense pressure from humans. Part of this is simply habitat degradation. Mankind alters environments to suit its own needs, taming the landscape to build farms and factories, and crisscrossing once remote wilderness with highways. But, sometimes we take much more direct action to rid the landscape of predators. Red wolves were once abundant throughout the southeastern portions of the US, but now only twenty or so survive. Today, in my home state of North Carolina, state agencies are actively trying to preserve the last wild population of red wolves. This is a change of heart; it was once the goal of governments to rid their territory of predators, and serious time and effort went into killing red wolves. We've made a last-second swerve in recent decades to rescue a species we spent three centuries trying to wipe out.

The story of the red wolf has its echo in the thylacine, a predator that was once found in Tasmania, New Zealand, and Australia. Thylacines are an unlikely member of the cryptid family. Cryptozoology is the study of unknown animals, but thylacines are very well known. Bits of bone and hide have been located in museums. There are numerous photos, and even film of them pacing back and forth inside their chain link cage in a zoo.

The creature revealed in photos is slightly unnerving to gaze upon. The back half of the beast is striped like a tiger, explaining why the species was commonly known as the Tasmanian tiger. Move forward on the body, and it stops looking like a tiger and starts resembling a wolf. But, it's not the blend of tiger and wolf that's most odd to the eye. My own gaze tends to focus on the face. The jaws are too long, the mouth opens too wide, the teeth are jagged, almost saw-like.

Thylacine at the London Zoo

In truth, the thylacine was neither a wolf nor a tiger. The thylacine is a marsupial, and it's more related to possums than it is to wolves or tigers. The similarity to other predators is a case of convergent evolution, where unrelated species take on similar body shapes to fill the similar ecological niches. The Tasmanian tiger is literally a textbook case for convergent evolution.

Alas, textbooks are the only place you find Tasmanian tigers now. The last known member of the species, a male thylacine named Benjamin, died on September 6, 1936, bringing a close to the story of a species that was once an apex predator.

The extinction of the thylacine was no accident. Tasmania offered bounties to anyone who brought in pelts. The thylacine was thought to be a threat to livestock, particularly sheep and chickens. Some people cynically believe government action never accomplishes anything useful. But, with over 2000 bounties paid, the Tasmanian government successfully protected Tasmanian chickens and sheep from thylacine predation. Mission accomplished!

Arguably, the species might have gone extinct without government intervention. Its range was limited, and it had already died out in Australia before Europeans began settling there. Extinction was the fate of nearly all species that existed before humans ever came along. The vast majority of extinctions have nothing to do with human predation or human caused environmental degradation, simply because humans are relative newcomers on the planet. Still, it's hard not to feel a twinge of guilt at the loss of the thylacine, a unique, magnificent beast, erased from the world, in part, by deliberate human malice.

Unless, maybe, we didn't get them all? Tasmania is over 26,000 square miles. Much of it is heavily forested, rugged terrain. Its population density is only 21 people per square mile, making it less densely populated than Nebraska. When bounty hunters went after Tasmanian tigers starting in 1888, the ones that were killed first were likely to be the boldest members of the species, the ones with the least instinctive fear of humans, blithely loping across open fields. The thylacines that lasted the longest against the bounty hunters were the most wary ones, the ones who hid in dense underbrush and sniffed the wind for human scents.

Could some small population of hyper-wary thylacines linger in Tasmania, or even perhaps in Australia? Thylacine cubs were cute. It's hard to believe that at least a few weren't smuggled off the island by Australian visitors.

Thylacine cubs

You could have thylacines lurking in Australian

wilderness for the same reason Burmese pythons now thrive in the Everglades.

We can speculate about possible paths to survival, but do we have any evidence? We run into a problem common to many cryptids. There have been numerous eyewitness sightings in recent decades. We have a few blurry, inconclusive photos and videos from trail cameras or cellphones. From time to time, footprints are found. Here the evidence stops. No thylacine scat has been found. No skulls of recent vintage. Not even a single tooth. With each year that passes, the odds that a breeding population can still be found in the wild grows ever smaller.

For some cryptids, scientists would demand a more or less intact corpse be produced to prove their existence. For a beast like the thylacine, known to exist until recent times, a clear, unmistakable photo might be enough to sway skeptics. The pathway to proving that living thylacines exist is much clearer than it would be for Bigfoot or Nessie. Not only do we know they existed, there are probably people still alive who witnessed those last few surviving in zoos.

What does the reality meter say?

Modern physical evidence: There have been possible footprints collected, but footprints point to the same problems we encounter with yeti prints. Find the prints of a deer or a bear, and you'll also likely find scat or hairs. At best, we can give this 2 points.

Fossil or other preserved evidence: No one disputes these animals were once real and relatively easy to find in the wild. We have their hides, bones, and DNA. A solid 20.

Biological viability: Nothing could prove that they were biologically viable better than the fact that we have bodies to study. Another 20.

Habitat: There's still a lot of wilderness in the previous range of the thylacine. But, humans have altered the terrain a great deal, and many of the species the thylacine evolved to feed upon are also endangered. It's plausible that an apex predator could still find enough food to support a breeding population, but hardly guaranteed. I think this deserves a 15.

Concealment: Tasmania houses many unique species found nowhere else in the world, so scientists spend time and effort studying even the most remote nooks and crannies of the island. With modern motion sensing trail cams, any remaining thylacines are simply running out of place to hide. That said, if thylacines are sensitive enough to human scents that they won't even go near a place where a human has been to set up a camera, it's vaguely plausible that we'll need better tech to find them. Even allowing for this, I can't grant more than 4 points here.

Alternate explanations: The most likely explanation for modern thylacine photos and videos is that they are stray dogs or cats. The reason every modern thylacine photo is blurry is simple: If the camera is in focus, it's obvious we're looking at a dog. On the other hand, dogs aren't supernatural beings. If a scientist were to swiftly investigate a potential thylacine sighting, they could probably find the stray dog to blame for the sighting. I feel like this falls somewhere in between near certainty and pure speculation, so I'll only deduct 10 points.

This brings the Tasmanian tiger to a final reality score of 51, almost perfectly balanced between reality and unreality. The modern thylacine isn't likely to be real, but there are still threads of evidence that lets a rational person hold out hope that one will turn up. As I write this, not even a century has passed since the last known specimen died. I find creatures like the yeti plausible based on bones and remains thousands of years old. If a thylacine is captured tomorrow, or is even clearly filmed, I'll not only accept it as real, I'll boast that I thought it was real all along.

The Reality Meter
Thylacines

Modern Physical Evidence	2
Fossil/Preserved Evidence	20
Biological Viability	20
Habitat Strength	15
Concealment	4
Alternative Explanations	-10

CHUPACABRA

Chupacabra

The chupacabra is a relative newcomer to the cryptid universe. While there were proto-chupacabra stories from Puerto Rico as early as the 1970s, the creature became famous overnight for the same reason the Abominable Snowman exploded in popularity—it got labeled with a catchy, evocative name.

The precise origin of the name seems well established. Stories of goats or other livestock being drained of blood through puncture wounds in their necks were in circulation in Puerto Rico in mid-nineties, and in 1995 a San Juan radio DJ named Silvio Perez made a joking reference to a creature called the chupacabras, which translates into English as "goat-sucker."

The term spread like wildfire, leaping across open water from the island of Puerto Rico to land on the larger territories of North and South America, leading to reported sightings ranging as far north as Maine and as far south as Chile. In the United States, the creature lost the "s" at the end of its name, probably because English-speakers think of words ending in "s" as plural. There's a curious rule of cryptids that, with a few exceptions, most of them are discussed as singular entities. Thus, we refer to "the Abominable Snowman," but almost never to "Abominable Snowmen." The Loch Ness Monster, Bigfoot, Mothman—almost always we speak of them as if there's only one specimen.

The chupacabra might have lost its original spelling but it found fame with blistering speed. Only two years after a radio DJ coined the term, the popular television show "X-Files" built an episode around a chupacabra investigation. Since then, the creature has appeared in documentaries, comic books, children's cartoons, and thousands of hours of YouTube videos.

The description of the chupacabra is somewhat fluid. Is it the size of a bear? Or only a little bigger than a dog? Does it have hair or scales? Horns? Hooves? Spikes along its spine? Does it have wings? Does it go about on two legs or four? Do its footprints look canine, reptilian, or human?

Chupacabra "corpses" occasionally turn up. Nearly all are dismissed by scientists as canines with mange. To the extent there is actual physical evidence of something killing goats or livestock, feral dogs likely explain a good deal. Dogs kill their prey by bites to the throat, which would explain the puncture wounds. Solitary dogs without a pack are likely to slink away unseen if it hears or smells humans coming near. A human hearing a commotion outside might walk out of his farmhouse to find one of his goats dead and no culprit in sight.

Dead livestock are common starting points for numerous cryptids and local lore. The story of intact bodies drained of blood gets blamed on everything from aliens, to secret government research, to Satanic rituals. Here in North Carolina, one of our homegrown cryptids is "the Beast of Bladenboro." The legend of the beast is built around a series of animal deaths in the 1950s where animals—frequently dogs—were found dead and mangled, slain by some unknown beast. I've no reason to doubt that violent animal deaths happened, or that the culprit made several kills without leaving behind definitive evidence of its identity. I also suspect that, if similar reports were to emerge today, some people would feel there was no mystery at all to the creature's identity. Obviously, there's a chupacabra in the neighborhood.

Reality meter time!

Modern physical evidence: I'll work on the assumption that farmers aren't slaughtering their own livestock to hoax people. Unknown animals do sometimes kill farm animals and family pets. The term "unknown" has a lot of wiggle room. A chupacabra can slide into this space with a generous score of 2.

Fossil or historical evidence: Here, the ambiguity of what the creature looks like thwarts us. We can't determine if any fossils resemble it if we can't establish a more firm agreement on what it looks like. I'll have to score this a 0.

Biological plausibility: We might not know what a chupacabra looks like, but we do, at least, have some clue to its diet. It seems to feast upon blood, shunning the meat. This isn't the most effective diet for a large animal, but mosquitoes thrive on a blood diet, as does the vampire bat, and even a bird—the vampire finch. Notably, none of these creatures directly kill their chosen sources of blood, while I'm unaware of the chupacabra leaving survivors. Since unrelated creatures have evolved to feast on blood, I'll again be generous and score the possibility of a large blood feasting creature as a 4.

Habitat: Without knowing whether the chupacabra is a mammal, a reptile, or some horrifying large variant of vampire finch, it's difficult to state firmly that any given habitat is a good match. On the other hand, if it can live anywhere there's livestock to feast upon, that's a reasonably large range. We'll give this a 5.

Concealment: Unfortunately, if it does feast on farm animals, the chupacabra must travel across land that humans live upon. Cryptids can plausibly hide in the deep ocean or in rugged wilderness far from any road. But to remain hidden in farmland, with open fields and pastures, feels like more of a stretch. Add in the fact that people in rural areas often have hunting dogs, and it's nearly impossible to see how a truly exotic creature can long elude detection. This is another 0.

Alternative explanations: The vast majority of chupacabra corpses turn out to be dogs. The small animal victims of the chupacabra are likely targets for dogs, and the wounds are consistent with how a dog kills prey. On the other hand, dogs aren't particularly difficult to identify. If a farmer does have a feral dog killing chickens or small goats, he's got a financial motivation to deal with that problem rather than making up stories of a mystery beast. I'll only deduct 10 points.

This brings the chupacabra's final score to a 1. The goat-sucker is the least grounded in reality of any cryptid in this book. The primary insight a cryptid hunter should come away with is that enduring legends can be built upon the thinnest possible evidence. An evocative name can capture the imagination even more than fuzzy photographs or the occasional footprint.

The Reality Meter
Chupacabra

Modern Physical Evidence	2
Fossil/Preserved Evidence	0
Biological Viability	4
Habitat Strength	5
Concealment	0
Alternative Explanations	-10

KRAKEN

Kraken

Release the kraken! The myth of a giant, unknown cephalopod larger than ocean-going boats was widespread enough to inspire appearances in the literary works of Melville, Tennyson, Hugo and Vernes. Woodcuts showing enormous octopuses with their tentacles entwining the masts of ships date back to the early 1800s. No doubt, many a sailor testified to having witnessed it with their own eyes.

Which, perhaps, is why giant cephalopods felt more like myth than reality for quite some time. Sailors of previous eras didn't collectively have the greatest reputation for unceasing devotion to reporting only the truth. A flair for colorful storytelling seems to have been a requirement for the job. Individual tales of mermaids were likely taken with a grain of salt. Why should the kraken be treated any more seriously?

Here, the normal script for cryptids gets flipped. Many cryptids cling to reality through eye-witness testimony and blurry photos, but no physical remains. But, the largest cephalopods, the unimaginatively named giant squid and colossal squid, have been known to scientists for at least two centuries because dismembered tentacles would wash ashore. Remains also turned up in the bellies of sperm whales. The hides of such whales were often scarred with lines of large circles, as if they'd been grappling with a creature possessing raspy suckers on enormous tentacles.

What scientists lacked was any recorded visual evidence showing the creatures in their natural habitat, or any direct observation of a living specimen. This fact is even more remarkable when you realize that the largest squid can grow longer than some species of whales. How hard can it be to find a creature that's one of the largest animals on the planet?

Really hard, it turns out. Whales can be observed without much effort because they need to surface to breathe. Supersized squid have little reason ever to rise to surface waters. Searches were conducted for many decades, but it wasn't until 2004 that a living giant squid was finally photographed in its deep ocean habitat. The colossal squid remains unphotographed. No aquarium to date has ever displayed a living specimen of either beast.

The reality of giant squids isn't a matter of any serious dispute. But, are giant squids the kraken of yore? The most famous woodcut of a giant cephalopod plainly shows a giant octopus, not a squid.

Woodcuts aren't digital cameras. I think it's fair that we allow for a bit of artistic interpretation. The mistakes Dürer made in illustrating a rhinoceros doesn't disprove the existence of rhinos. Since we have physical proof that giant squids exist, and squids and octopus are both  tentacled creatures with large eyes and a mantle, the probability that kraken sightings were actually giant squid sightings strikes me as reasonably high. The largest colossal squid is estimated to be a little more than 40 feet long. In the modern world, where we have cruise and cargo ships hundreds of feet long, 40 feet feels a little underwhelming. But, one of the ships Columbus used to cross the Atlantic, the Nina, was only 50 feet long. If you were a crew member on such a vessel and a creature nearly as long as your ship swam up from the depths, it likely made quite an impression. A colossal squid is twice as long as a giraffe is tall. I think it can be called monstrous, at least in appearance.

As for monstrous behavior, squids don't aggressively attack ships. But if an enormous squid were accidently hooked on an anchor or harpooned like a whale, it's plausible to imagine it whipping its tentacles around in an attempt to defend itself.

If we do accept that the kraken was a colossal squid, it leaves the cryptid hunter in a curious position. The dream of any cryptid hunter is to prove the existence of a creature that's been reported, but never proven to exist. But success comes with a sort of bittersweet reward. Once it's confirmed by science, a cryptid stops being cryptic. It's now part of a known species studied by reputable scientists who reject the label of cryptid. But, returning to the root of the word, cryptozoology translates as the study of hidden animals. Despite a few videos and the occasional corpse, these monsters of the deep continue to remain hidden. So, even though we know it's real, I would argue it's still a proud member of the cryptid club.

Time for the reality meter!

Modern physical evidence: We have bodies! We have video! 20 points!

Fossil and historical evidence: Relatives of squid are very common in the fossil record. Some may have been even larger than today's specimens. Another 20!

Biological viability: Cephalopods are abundant throughout the oceans, and have a run on this planet that puts dinosaurs to shame. 20!

Habitat: Oceans cover 70% of the world's surface, and the tallest mountains would be swallowed by the deepest ocean trenches. Kraken have plenty of elbow room, despite lacking actual elbows. 20!

Concealment: There's roughly a 200 year gap between naturalists getting to examine a colossal squid tentacle and scientists filming one in the wild. That's some world championship hide and seek talent. An easy 20.

Alternative explanations: Currently, the kraken has a perfect score. But, what if my assumption that the colossal squid inspired the kraken is mistaken? Might there yet be an unknown cephalopod that's even larger still hiding in the depths? Given the stunning effort it took to film a living giant squid, if an even larger species turned up I wouldn't be shocked. But, there's also the possibility that initial stories of the kraken weren't based on a real creature at all. They might have been tall tales told for amusement. It was only coincidence that an actual species was discovered that shared some of the elements of the myth, similar to the way the Komodo dragon kinda, sorta looks like a dragon. Because we lack a time machine to go back and interrogate long dead sailors in an attempt to clarify their accounts, we must at least accept there's a chance that the kraken was something other than a giant squid. I'll deduct a single point, since this is mere speculation. If you're a cryptid hunter convinced that the true kraken is still waiting to be found, I want to leave you some wiggle room. -1.

Final score: A commanding 99. Among cryptids, the kraken is the reality champion.

The Reality Meter
Kraken

Modern Physical Evidence	20
Fossil/Preserved Evidence	20
Biological Viability	20
Habitat Strength	20
Concealment	20
Alternative Explanations	-1

The Most Important Search

Before I close out this book, I wanted to take time to address two theories that turn up in the study of cryptids. I've addressed the existence of various cryptids doing my best to ground these creatures in natural law. Whenever possible, I've tried to place them into their appropriate branch of the evolutionary tree, since all life on Earth is connected to other living things.

But what if creatures like the chupacabra or Mothman aren't originally inhabitants of Earth? The universe is vast, and the processes that created life on Earth arise from basic physical laws that apply across galaxies. Gravity leads to the formation of stars, and the dust around stars gets formed into planets. The ratios will vary, but the elements that make up these alien worlds are the same elements that form our own world—iron, carbon, silicon, oxygen, and so on. Even if the odds are a slim billion to one that a planet might host life, the number of stars, and the number of potential planets, is so immense that there must be millions of living worlds, and among these worlds there must be numerous civilizations.

I believe in alien civilizations as a statistical likelihood, but also will gladly admit that believing in alien life is ultimately an act of faith, for now. We simply have no clear evidence that they are out there.

We have even less evidence that they are here. It's not something that can be categorically ruled out. There's a good argument for why aliens must exist. But, good arguments can't substitute for evidence. Right now, I have evidence for dinosaurs and extinct hominids and giant birds and huge beasts of the sea. Until alien life capable of visiting our planet is verified, it seems unfruitful to invoke aliens as an explanation for cryptids.

There's a second all-purpose explanation that is sometimes invoked. What if dragons, mermaids, and thunderbirds aren't physically real, but are instead supernatural entities that we simply can't understand through the laws of science? What if we're glimpsing creatures that, in previous eras, would be described as demons or angels, or perhaps as ghosts, or even gods?

It's beyond the scope of this book to argue the existence or nonexistence of the supernatural. If such explanations appeal to you, I see no reason to try to dissuade you. But, if cryptozoology is to present itself as a serious attempt at the study of real creatures, it should confine itself to the same rules that apply to zoology. We need to be hunting for living, breathing beasts with physical bodies.

In closing, I want to toss out one more idea. It would be satisfying if every cryptid discussed in this book were one day proven to exist. But, for most of them, for any number of reasons, we just will never know. We only get to witness a small slice of reality, a limited portion of both space and time. Our brains evolved primarily to find food and mates. It's the best tool we have, but it's still a somewhat crude instrument for grasping physics and biology.

The world has existed for several billion years before our birth and will continue spinning several billion years after we're gone. We simply can't know everything that came before, we'll never know what's going to come after, and we're blind to most of what is happening in our world right now. We pay no attention to mites feasting on our dead cells in our eyebrows. We think nothing about the bacteria thriving in our guts. The giant squid swims through deep, dark seas surrounded by things we'll never see, and can't imagine.

In this light, the fact that anyone would claim that we've already discovered every large creature in the wild is a foolish claim. Just because something is unknown isn't proof that it's unreal. If someone tells you a certain cryptid doesn't exist, there's no reason to accept that as the final word. There can be no final word. Reality is still being discovered and understood, and there's an infinite supply of the unknown to transform into the knowable.

Go search.

Go out into the forests and swamps, go up mountains, and paddle over remote lakes, forever searching. You might find Bigfoot or Nessie. Probably not. But you will definitely discover small slices of the here and the now, and come to better understand your world and your relationship to it.

Go searching for cryptids, and you might just find yourself.

ABOUT THE AUTHOR

JAMES MAXEY'S mother warned him if he read too many comic books, they'd warp his mind. She was right! James is unsuited for decent work and ekes out a pittance writing down demented fantasies about masked women, fiery dragons, and monkeys.

Readers interested in sampling Maxey's odd ramblings might enjoy his science-fantasy *Bitterwood* series, the D&D inspired fantasy of his *Dragon Apocalypse* novels, his two superhero series *Lawless* and *Whoosh! Bam! Pow!* or the steam-punk visions of *Bad Wizard*. His short fiction has appeared in *IGMS*, *Asimov's*, and over a dozen anthologies, with the best of his work appearing in the collections *There is No Wheel* and *The Jagged Gate.*

James lives in Hillsborough, North Carolina with his lovely and patient wife Cheryl and too many cats. To sign up for his newsletter, visit jamesmaxey.net. He can be found on Facebook by searching for the group Dragonsgate: The Worlds of James Maxey. Or, follow him on Twitter @JamesAllenMaxey

Links!

Share your feelings about this book with a review! Just hover your smartphone camera over this image to be taken to the Amazon page for this book! Even just a rating is helpful!

Use this link to find other books on James Maxey's Amazon Author Page